Advances in

ECOLOGICAL RESEARCH

VOLUME 35

Advances in Ecological Research

Series Editor: H. CASWELL

Biology Department
Woods Hole Oceanographic Institution, USA

Advances in

ECOLOGICAL RESEARCH

VOLUME 35

Birds and Climate Change

Edited by

ANDERS P. MØLLER

Université Pierre et Marie Curie
Laboratoire de Parasitologie Evolutive
Paris, France

W. FIEDLER AND P. BERTHOLD

Max Planck Research Centre for Ornithology
Radolfzell, Germany

2004

ELSEVIER
ACADEMIC
PRESS

Amsterdam   Boston   Heidelberg   London   New York   Oxford   Paris
San Diego   San Francisco   Singapore   Sydney   Tokyo

ELSEVIER B.V.	ELSEVIER Inc.	**ELSEVIER Ltd**	ELSEVIER Ltd
Radarweg 29	525 B Street, Suite 1900	**The Boulevard, Langford Lane**	84 Theobalds Road
P.O. Box 211, 1000 AE Amsterdam	San Diego, CA 92101-4495	**Kidlington, Oxford OX5 1GB**	London WC1X 8RR
The Netherlands	USA	**UK**	UK

First edition 2004

ISBN: 0-12-013935-9
ISSN: 0065-2504 (Series)

⊗ The paper used in this publication meets the requirements of ANSI/NISO Z39.48-1992 (Permanence of Paper). Printed in The Netherlands.

Working together to grow libraries in developing countries

www.elsevier.com | www.bookaid.org | www.sabre.org

ELSEVIER   BOOK AID International   Sabre Foundation

Contributors to Volume 35

FRANZ BAIRLEIN, *Institute of Avian Research 'Vogelwarte, Helgoland' An der Vogelwarte 21, D-26386 Wilhelmshaven, Germany*

PETER BERTHOLD, *Max Planck Institute for Ornithology, Vogelwarte Radolfzell, Schlossallee 2, D-78315 Radolfzell, Germany*

KATRIN BÖHNING-GAESE, *Institute for Zoology, Dept. V-Ecology, Johannes Gutenberg – University Mainz, Becherweg 13, D-55099 Mainz, Germany*

CHRISTIAAN BOTH, *Netherlands Institute of Ecology, PO Box 40, Boterhoeksestraat 22, NL-6666 ZG Heteren, Netherlands*

TIMOTHY COPPACK, *Institute of Avian Research 'Vogelwarte Helgoland', Inselstation, D-27494 Heligoland, Germany*

PETER DUNN, *Department of Biological Sciences, University of Wisconsin-Milwaukee, PO Box 413, Milwaukee, WI 53201, USA*

STEINAR ENGEN, *Department of Mathematical Sciences, Norwegian University of Science and Technology, N-7491 Trondheim, Norway*

WOLFGANG FIEDLER, *Max Planck Institute for Ornithology, Vogelwarte Radolfzell, Schlossallee 2, D-78315 Radolfzell, Germany*

OMMO HÜPPOP, *Institute of Avian Research 'Vogelwarte Helgoland', Inselstation, D-27494 Heligoland, Germany*

ULRICH KÖPPEN, *Hiddensee Bird Ringing Centre, LUNG Mecklenburg-Vorpommern, Badenstr. 18, D-18439 Stralsund, Germany*

MARCEL M. LAMBRECHTS, *CNRS/CEFE, 1919 Route de Mende, F-34293 Montpellier Cedex 5, France*

ESA LEHIKOINEN, *Department of Biology, University of Turku, FIN-20014 TURKU, Finland*

NICOLE LEMOINE, *Institute for Zoology, Dept. V-Ecology, Johannes Gutenberg – University Mainz, Becherweg 13, D-55099 Mainz, Germany*

JUHA MERILÄ, *Ecological Genetics Research Unit, Department of Ecology and Systematics, PO Box 65, University of Helsinki, FIN-00014 Helsinki, Finland*

ANDERS P. MØLLER, *Laboratoire de Parasitologie Evolutive, CNRS UMR 7103, Université Pierre et Marie Curie, Bat. A. 7 èmè étage, 7 quai St. Bernard, Case 237, F-75252 Paris Cedex 05, France*

FRANCISCO PULIDO, *Max Planck Institute for Ornithology, Vogelwarte Radolfzell, Schlossallee 2, D-78315 Radolfzell, Germany*

BERNT-ERIK SÆTHER, *Department of Biology, Norwegian University of Science and Technology, Realfagsbygget, N-7491 Trondheim, Norway*

TIM H. SPARKS, *NERC Centre for Ecology and Hydrology, Monks Wood, Abbots Ripton, Huntingdon, Cambridgeshire PE28 2LS, United Kingdom*

WILLIAM J. SUTHERLAND, *Center for Ecology, Evolution, and Conservation, School of Biological Sciences, University of East Anglia, Norwich NR4 7TJ, United Kingdom*

MARCEL E. VISSER, *Netherlands Institute of Ecology, PO Box 40, NL-6666 ZG Heteren, Netherlands*

MECISLOVAS ZALAKEVICIUS, *Institute of Ecology, Akademijos 2, LT-2600 Vilnius, Lithuania*

Preface

The climate—the average state of the atmosphere over a particular region, and the sequence of weather events characteristic of that part of the world—not only varies widely in the different climatic belts around the earth from equator to poles, but has also repeatedly undergone more or less pronounced changes in the course of time. Evidence of large global climate alterations in the earth's geological past is provided, e.g., by Tertiary plant remains in the Arctic and the presence of Jurassic coal in the Antarctic. In the European region, climate fluctuations during the Pleistocene, with a rapid alternation between warm and cold periods, have had a long-term significance, especially in view of the fact that the last Ice Age peaked less than 20,000 years ago, and the fauna and flora typical of the present day evolved thereafter, during the last 10,000 years. As the study of climate developed into a science—climatology—the period from 1931–1960 became defined as the "normal period", in comparison with which it is ever more obvious that we are currently experiencing a progressive increase in world temperature: a "period of global warming". However, many people—among them many scientists—are still sceptical and uncertain as to how the evidence of recent climatic warming and its possible consequences should be evaluated. There are three main reasons for this scepticism and uncertainty. First, even a clearly evident climatic change does not proceed without interruption; the weather is capricious, and even a single unusual cold spell (like that in Europe during December 2001) can be regarded by sceptics as "proof" that there are no grounds for speaking of a general global warming. Secondly, the short- and longer-term weather conditions are so changeable that an individual cannot, from personal experience and memory, compare the weather in successive years so accurately as to draw objective conclusions about climatic change. Systematic changes in climate can only be demonstrated to us by meteorologists, on the basis of many years of careful, large-scale measurements and comparisons of average values; but these assessments are not easy for many of the public to understand. Thirdly and finally, many people do not make a clear distinction between the moderate warming that *has already happened* in the last 100 years or so—and which is beyond dispute—and a *predicted greater warming in future*, inferred from very recent experience and currently the subject of vigorous debate.

In response to the marked global warming above all over the past decades, in 1998 the UN established a special investigative committee, the "Intergovernmental Panel on Climate Change" (IPCC). This international group of experts is examining the causes of the present alterations in climate, periodically publishes "best estimates" describing realistic expectations for future climate changes, and offers advice to governments and the like. In the view of the IPCC not only has the greatest warming within the last 1000 years already occurred, but the estimates for a further temperature rise by the year 2100 vary between about 1.5–5.5°C, ascribable to pollution of the atmosphere with so-called greenhouse gases.

Birds, as organisms with a very active metabolism, are highly sensitive to climatic changes, and as highly mobile creatures they are also extremely reactive. Thus, for some decades now the birds of many parts of the world have been exhibiting clear changes in a number of respects, often similarly directed in different species. These changes are so pronounced that, even if we knew nothing about the simultaneous climatic warming, we would be forced to conclude that some environmental factors have been undergoing systematic alterations that encouraged many bird species, for example, to expand their ranges to higher latitudes, to return earlier to breeding grounds, to stay there longer, to prolong breeding periods etc.

Because the biology of birds has been more thoroughly investigated than that of any other group of organisms, birds were pioneer indicators for changes related to global warming; as early as 15 years ago it was possible to make prognoses for them (Berthold, 1991: Patterns of avian migration in light of current global "greenhouse" effects: A central European perspective. Acta XX Congr. Int. Orn., Christchurch 1990, 780–786). Since then such a large amount of data on birds has accumulated that it was urgently needed to be discussed at a special symposium, and the present proceedings are the result. It is to be hoped that these reports will contribute towards keeping birds in their pioneer role for the ongoing analysis of changes and adaptations in a time of rapid climate change—an opportunity that Charles Darwin would probably have envied. We would like to thank all the contributors for their collaboration. Numerous scientists are gratefully acknowledged for providing reviews of the chapters. The European Science Foundation and Prof. F. Bairlein in particular provided funding and help with organising the workshop on birds and climate change. Finally, we would like to thank the editors from Academic Press, J. Meyer and S. Mitra, for their help and encouragement.

Peter Berthold
Anders P. Møller
Wolfgang Fiedler

Acknowledgements

The chapters in this volume are largely based on contributions to a workshop "Bird Migration in Relation to Climate Change" that was organized in March 2003 by the Laboratoire de Parasitologie Evolutive, CNRS, Paris, and the Max Planck Institute for Ornithology, Vogelwarte Radolfzell, and hosted by the University of Constance.

We are grateful to the European Science Foundation (ESF) for funding this Workshop.

Contents

Arrival and Departure Dates

ESA LEHIKOINEN, TIM H. SPARKS and MECISLOVAS ZALAKEVICIUS

Migratory Fuelling and Global Climate Change

FRANZ BAIRLEIN and OMMO HÜPPOP

Using Large-Scale Data from Ringed Birds for the Investigation of Effects of Climate Change on Migrating Birds: Pitfalls and Prospects

WOLFGANG FIEDLER, FRANZ BAIRLEIN and ULRICH KÖPPEN

Breeding Dates and Reproductive Performance

PETER DUNN

Global Climate Change Leads to
Mistimed Avian Reproduction

MARCEL E. VISSER, CHRISTIAAN BOTH and
MARCEL M. LAMBRECHTS

Analysis and Interpretation of Long-Term Studies
Investigating Responses to Climate Change

ANDERS P. MØLLER and JUHA MERILÄ

Photoperiodic Response and the Adaptability of Avian Life Cycles to Environmental Change

TIMOTHY COPPACK and FRANCISCO PULIDO

Microevolutionary Response to Climatic Change

FRANCISCO PULIDO and PETER BERTHOLD

Climate Influences on Avian Population Dynamics

BERNT-ERIK SÆTHER, WILLIAM J. SUTHERLAND and STEINAR ENGEN

Importance of Climate Change for the Ranges, Communities and Conservation of Birds

KATRIN BÖHNING-GAESE and NICOLE LEMOINE

The Challenge of Future Research on Climate Change and Avian Biology

ANDERS P. MØLLER, PETER BERTHOLD and
WOLFGANG FIEDLER

Arrival and Departure Dates

ESA LEHIKOINEN,* TIM H. SPARKS AND
MECISLOVAS ZALAKEVICIUS

I. SUMMARY

We review the current evidence that changes in arrival and departure dates of migratory birds have taken place, and the relationship of these changes to climate variability. There is little doubt that the timing of spring migration closely follows weather variations. This is more evident and/or stronger in short-distance migrants than in long-distance migrants, but the latter have also responded to climate change. Changes of spring arrival in birds depend on climate impacts at different latitudes along the route from wintering to breeding areas. Therefore, also migratory strategy, e.g., stopover tactics and migratory routes, may be under selective forces due to climate change. Changes in breeding environment depend on the climate there. The discrepancy between *en route* and breeding time impacts can induce poor fit between annual cycles of birds and their resources.

E-mail address: esalehi@utu.fi (E. Lehikoinen)

ADVANCES IN ECOLOGICAL RESEARCH, VOL. 35
0065-2504/04 $35.00 DOI 10.1016/S0065-2504(04)35001-4

Changes of timing of autumn migration are more variable and they are less well understood. Some species have advanced and others postponed their autumn migration. As long as seasonal variation of environmental constraints at species level remains elusive we cannot predict which species will delay autumn departure and which will advance it in synchrony with spring events.

The net result of changes of phenology has often been the lengthening of the summer part of the annual cycle. Because the first analyses of changes of both migration periods and time span between them are based on bird station data, it is still too early to generalise this variation against ecological traits of species. Lengthening of time spent in the breeding area may relax some time constraints set by seasonality by allowing more time to breed and moult, but just the opposite is also possible depending on latitude and temperature regime at which changes are taking place.

II. A BRIEF HISTORY

The earliest systematic records on arrival and departure of migratory birds in Great Britain are from the early 1700s (Kington, 1988; Sparks and Carey, 1995) and in Northern Europe from the mid-1700s (Linné, 1757; Leche, 1763). These texts mark the start of phenological research, which was one of the earliest organised branches of systematic ornithological observation in many parts of Europe.

It has been well known for hundreds of years that arrivals of migrating birds are good predictors of advancing spring. In order to test their potential usefulness for making decisions on when to start certain agricultural activities in the mid-18th century, systematic phenological observations were encouraged, e.g., in Sweden (including Finland at the time) by A. Celsius and C. Linné. The purpose was commercial—to aid the country's economy. Johannes Leche, professor of medicine in the Academy of Turku and a friend to Linné, did not do a regression analysis like that shown in Figure 1 based on his data (Leche, 1763), but he phrased the essential information in the following words.

The White Wagtail is always observed at the sea ice break. If it is sometimes observed earlier, the ice will break very soon. Of all hirundines, the House Martin is the first to arrive. This species cements its nest so tight that it can hardly enter it herself. On average, it arrives on May 6th. The Barn Swallow, the species with a brownish patch under the bill, arrives on May 10th. When these species arrive, the temperature of water in the bays is 9 or 10°C. He who has not yet started to take care of his herb garden should not postpone it any further. The Common Swift is the last to arrive, sometimes before, sometimes after May 20th. At this time, the real summer begins. You can now sow the purslane (*Portulaca oleracea sativa*) and plant the cucumber even on cold soil, as well as the beans, without a fear of frost.

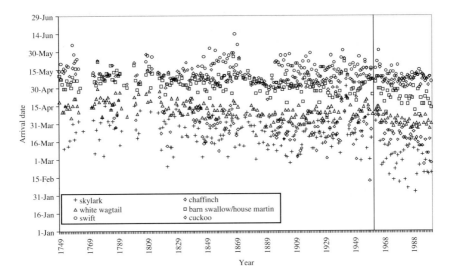

Figure 1 Dependence of arrival date of selected species on average temperature of the previous month in Turku, SW Finland in 1749–1762. The regressions of Wagtail ($b = -2.34$ days/°C $n = 14$) and Swallow ($b = -2.01$, $n = 15$) are statistically significant while that of the Swift ($b = -1.02$, $n = 13$) is not.

Since the mid-18th century, the importance of long term phenological data collection has been reinvented several times and in many countries. Thanks to this, we now have in Europe almost continuous time series of arrival and departure dates for two centuries or more (e.g., Finnish Society of Science and Letters' time series from 1749; e.g., von Haartman, 1956; British Marsham data, Sparks and Carey, 1995). These time series provide an opportunity to look at arrival–weather relationships in different situations. This is a benefit, because analyses of only recent data during a period of predominantly warming temperature, in those areas where most data are available, does not provide us with the other "experimental" group: the response of birds against climate becoming progressively colder. A few of these older data series give an opportunity to look at this situation.

We will first have a short look at the history of bird phenology. Then we describe how timing of migration can change. We will then look at the quality and types of data available and response variables and evaluate the methods. Finally, we summarise what kinds of responses to climate change have been observed in different areas and species. A preliminary meta-analysis will summarise the information available to us both as trends and as responses to weather factors.

III. VARIATION OF ARRIVAL DATES DURING THE LAST 250 YEARS

The longest continuous time series available for analysis exceed 250 years in length. All these data are observations of first arrivals of bird species that were well known to people of rural communities. In time series approaching or exceeding 100 years (Sparks and Carey, 1995; Ahas, 1999), it becomes readily evident that the variation of arrival times around the overall average or trend often shows periods of constant direction of change and then a shift in the direction (Figure 2). Correlating or regressing the arrival dates with a suitable weather (often temperature) variable often reveals a strong relationship between arrival and temperature, but a substantial part of the variation remains unexplained (Figure 1). A proportion of unexplained variation arises from the fact that the local temperature that is used in correlation and regression analyses is, at best, only a proxy of the weather along the migration route prior to arrival at the observation locality.

These records also show that most, if not all, migratory bird species react behaviourally to spring weather; arriving early in warm springs and late in cold springs. This is not unexpected. Every amateur ornithologist learns this by the second year of his hobby. The methodological message provided by long time series is that we should be cautious when short recent time series are used, because of varying directions of "trends".

The use of the longest time series of first arrivals in interpreting the variation in arrival times is, however, limited for a few reasons. First, we have little information of other factors that might have caused changes to the arrival time. We usually have only a rather weak idea of habitat variability over long periods and especially of changes of population size (Sagarin, 2002). Nor do we know accurately how observer activity has varied over time. Despite these uncertainties, several of the long time series give a rather credible picture of changes in migration timing in the past.

IV. CURRENT DATA AND APPROACHES

A. Types and Quality of Data

Different types of arrival information have been used in recent analyses. First arrivals and mean or median arrivals are the most common ones. *First arrival* (the day on which the first individual of the species is recorded each spring in the area) is most frequently recorded in some countries by local ornithological clubs and in others more systematically by meteorological stations or in programs run by scientific or governmental organisations

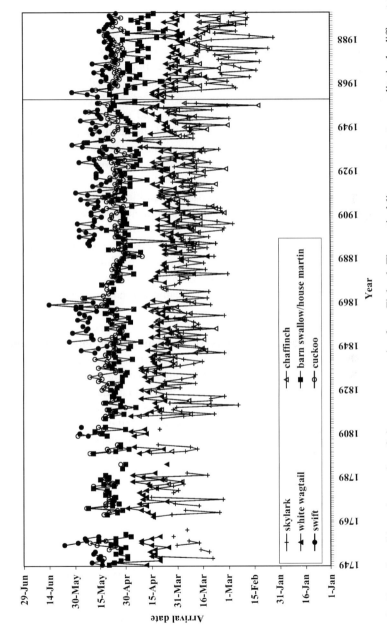

Figure 2 Examples of long time series of six common species from Finland. The vertical line separates two qualitatively different time periods—the data collected by the Finnish Society of Science and Letters (until 1965) and the modern data collected by the Ornithological Society of Turku (from 1965).

5

(von Haartman, 1956; Ahas, 1999; Sparks and Braslavská, 2001). Some first arrival data used are based on more or less repeatable observation practice (standard trapping, e.g., Hüppop and Hüppop, 2003 or observation, e.g., Zalakevicius, 2001a,b), but many include an unmeasured component of variation caused by variation of activity. First arrival dates do not necessarily reflect the migration behaviour of the whole population. *Mean arrival* (or median arrival) of a breeding population can be recorded in intensive studies. It is seldom done separately from the start of breeding. *Mean migration time* can be recorded in bird stations and similar sites, where all observed or trapped individuals passing the site are counted in a standard way. The difference of *mean arrival* and *mean migration time* has not been explicit in studies of the timing of bird arrival. It is, however, necessary to make this distinction, since migration data nearly always are a mixture of many different populations from only partially identified areas. The timing of annual cycles of these populations is under the influence of different breeding and wintering conditions and may therefore differ. Further, weather effects on migration increase error variance and thereby decrease chances to observe real effects. Yet, the bird station data are worth detailed analysis especially because they offer the possibility of studying the ways in which the whole dynamics of spring migration respond to weather variation and changes in climate as well as other factors.

First arrival, mean arrival and mean migration time do not necessarily measure the same response. The benefits and weaknesses of all point estimates of arrival distribution should be known. We list some properties of these measures in Table 1. Sparks *et al.* (2001) compared different point estimates of arrival distribution and observed that the variability of arrival date decreases from first to median bird. This pattern appears to lead to stronger temperature and weather variable correlations with early time points. For example, in Vähätalo *et al.* (2004) the number of significant negative correlations decreased from first individual, through 5% point estimate to median and 95% point of arrival distributions. This is probably partially a consequence of different individual responses to weather although the numbers of migrant individuals also affect the point estimates and correlations.

The questions of data quality and sampling problems are important when using measurements that were originally collected either as a hobby or for a purpose other than the current one. We believe that first arrival observations of long time series were often of the highest quality possible at the time of recording. For example, in Finland the two most extensive early studies were undertaken by the University of Turku (Leche, 1763; Justander, 1786) and by the Finnish Society of Science and Letters (a research programme started in 1846 and continuing today). The latter was carefully planned and a 70-page guide to describe observational methods was distributed (Nervander, 1846).

Table 1 Typical response variables used to describe arrival/departure times of migratory species

Variable	Definition	Problems	Benefits
First arrival/last departure	First individual observed in spring/last in autumn (of unknown status, local breeder or transient migrant)	Large random variance; atypical behaviour; data quality tests mostly lacking	Easy to observe, cheap, volume of data
Median arrival/departure	The middle individual arriving in/departing from a closely followed *breeding* population	Difficult, labour-intensive, requires special study	Closest to fitness consequences
Mean arrival/departure	Average arrival/departure date of all birds followed in the study area	As above, but not as easy, because of complex arrival distributions	Close to fitness consequences
Median/mean migration time	The middle or average date of migration in an intensively studied migration flow	Unknown mixture of passing populations, difficult statistical distributions, problems with mixing breeding populations	Is done in bird stations in standard ways, plenty of data available

The first bird stations started to operate around the beginning of the 20th century as a response to the main goal of the first International Ornithological Congress in Vienna 1884. It took, however, decades to develop the activity at the stations to the level that the whole dynamics and mean migration time could be determined reliably. Until the mid-1900s, we therefore usually have only first arrivals to use. This is a pity, because the climate warming period of 1880–1900 to 1930 offered an interesting comparative research period. Many bird stations standardised their methods of recording migration or trapping especially from the 1960s on. Thanks to this, data from bird stations can be used for studying the changes of migration patterns over the last three to five decades.

Bird stations are not often destinations for migrating birds. Rather, stations are situated on the migration route. This is technically a good thing, since the migration pattern can be analysed without the disturbing effect of local breeders inflating daily numbers counted (Bairlein and Winkel, 2001; Hüppop and Hüppop, 2003), but biologically this complicates things. While the first arrival records made in a given observation area and first arrival/ringing observation at the nearby bird station are very likely measuring the same response, it is not clear that mean migration time at the bird station describes the mean arrival of this area's population. There is no simple way available to overcome this difficulty. One possibility is to compare migration times at bird stations with start of breeding in the supposed breeding area of migrants. In pied flycatchers in SW Finland the timing of breeding (first egg) on the mainland and median migration at a nearby bird station correlated highly significantly ($r = 0.61$, $p = 0.0004$, df = 27; Sippola, 2003).

So far, most research on timing of arrival has only looked at the above simple measures of response. First arrival measures the response of birds at one tail of a distribution (Loxton et al., 1998; Loxton and Sparks, 1999). The strongest sceptical view states that the "earliest" observations are, to a large degree, caused by "atypical behaviour" for external or internal reasons (Loxton and Sparks, 1999) and therefore this type of data tells little about the true response of the population. In recent bird station data with complete monitoring during the migration period, we can use first arrival and migration dynamics data together. This provides an opportunity to test how well first arrival data and migration dynamics data correlate with each other (Vähätalo et al., 2004; Sparks et al., in press; Tøttrup, 2004). Generally, there is a significant positive correlation between first arrival and later point estimates of the proportion of migration that has passed the observation area. Even in this favourable case, first arrival observation is only a single point in a migration distribution. It only tells how this end of the distribution responds to weather or climate.

It is most likely that migration behaviour should also be modelled as an individual response. The simplest models view seasonal migration dynamics as a single distribution with peak numbers somewhere between the start and the end. Although the distribution may be in practice, for many internal (e.g., population, sex and age class differences in timing) and external (e.g., weather) reasons, an approximation to a complicated mixture of many mathematical distributions, we can calculate points of cumulative percentages of birds, which have passed the observation site or arrived to the target area. These point estimates can be used as a set of response variables in analysing arrival variations within the migrating population. So far, median migration time and median arrival dates have been studied, but other percentile point estimates have already been examined as well (Vähätalo *et al.*, 2004).

V. HOW IS EARLIER ARRIVAL ACHIEVED?

For a migratory species, breeding area is by definition different from the wintering area. Therefore, birds cannot respond *behaviourally* to amelioration of climate in the breeding area unless the climatic conditions in the wintering area and/or in the area through which the species migrate are positively correlated. The longer the migration journey, the less likely high positive correlations in climate between areas crossed will be. Long-distance migrants, furthermore, have to pass over partially independent climatic zones. Behavioural responses may be direct or indirect or both. The phenological situation in the environment of the target area of each migratory step is also under climatic regulation. Climate warming creates opportunities for earlier arrival in the target area through earlier and improved food availability (Harrington *et al.*, 1999; Walther *et al.*, 2002). Indirect effects are probably most important, but there is no reason to discard the direct effects, which operate through easier progression of migration in favourable weather situations and relaxing maintenance costs.

In addition, *evolutionary selective response* of the timing of migration to climate change in the breeding area is likely, if earlier arrival is correlated with improved breeding success. Looking only at the arrival/departure data can therefore tell only half of the story. In predicting how species with different migration strategies respond to climate change, we must look at both types of response—behavioural (or proximate) and selectional (or ultimate) (see, e.g., Coppack and Both, 2002). We leave the evolutionary response mainly as a task of chapters "Microevolutionary Response to Climatic Change; Climate Influences on Avian Population Dynamics; Importance of Climate Change for the Ranges, Communities and Conservation of Birds" and concentrate on looking at the arrival and departure changes and behavioural responses.

The three main ways to arrive earlier at the breeding area are (1) the wintering area has moved closer to the study area, (2) migration speed has increased and (3) migration starts earlier (see also Fiedler, 2001). Any combination of these is possible and when looking only at local data we cannot separate between them. Additional information is available for looking at cause (1), since ring recovery analysis might reveal changes in wintering areas, provided that the many other factors affecting recovery distributions can be accounted for (Fiedler *et al.*, 2004, this volume). For revealing the role of cause (3) we could in some cases make a study of departure from wintering areas and arrival in the breeding area where there is enough information to map the wintering area of known populations. There is very little published information on the dependence of departure from wintering area on weather conditions for any type of bird species: examples of non-significant effect of temperature exist for the American woodcock (*Scolopax minor*) in North America (Krementz *et al.*, 1994) and for the spotted flycatcher (*Muscicapa striata*) in South Africa (Kok *et al.*, 1991). Based on experimental evidence, it is widely accepted that the circadian rhythm, timed by photoperiod, controls departure at least from tropical wintering areas (Gwinner, 1996). Cause (2) is the hardest one to study, because of low numbers of useful recoveries and the inexactness and high variance in the calculation of migration speed. Its likelihood is increased by many recent studies, which strongly support the idea that tail winds may speed up migration (Richardson, 1990; Zalakevicius, 1993; Zalakevicius *et al.*, 1995; Liechti and Bruderer, 1998; Åkesson and Hedenström, 2000), but also by a very old one (Frederick II, of Hohenstaufen 1244–48). Frederick II was in fact very exact when stating, "They usually are able to choose a period of mild and favouring winds. North winds [in autumn] either lateral or from the rear are favourable, and they wait for them with the same sagacity that sailors exhibit when at sea." (From English translation of 1943 by Wood and Fyfe.)

Individuals very likely vary in the way they respond to climate change. Therefore, our response variables are a sum of all individuals' responses in the "population" passing the study area. Differences between individuals are reflected in the position, form and spread of the total migration distribution and the time point estimates (i.e., of first arrival, 5%, 50%, 95% and last bird, e.g., Vähätalo *et al.*, 2004, 1st, 3rd, 10th and median bird as in Sparks *et al.*, 2001) of the whole migration need not behave similarly in relation to climate variables. For example, Ptaszyk *et al.* (2003) have demonstrated change in the symmetry of the arrival distribution of white stork *Ciconia ciconia*. To make the scenario simple, we may visualise two extreme situations: (1) If only a small number of early migrating individuals are sensitive to warming the spread of the migration distribution increases and the response to temperature is seen only in those response variables describing the start of the migration of the species. (2) If all individuals are similarly sensitive, the date of the migration distribution changes without a change of spread and all time point estimates correlate with climate warming indicators. This clear situation, however, is complicated by an additional temperature impact. In many species, more than 50% of individuals arriving from spring migration are settling on their breeding site for the first time and in most of them young birds are not very faithful to their natal site, but rather strongly responsive to spring weather (Helbig, 2001). This causes additional variation in the migration distribution and may have important consequences for local population sizes. Much of the plasticity of birds in the phase of climate change may be realised through low natal site tenacity of young birds!

VI. STATISTICS

Several simple methods have been used to test whether changes in timing of migration are statistically significant. Some authors have used correlation (arrival vs. year, Pearson or Spearman) others have used linear regression (arrival vs. year) and in still other cases non-parametric smoothing methods have been used for descriptive purposes. In a few cases, multiple regression approaches have also been applied.

The statistical complications, of which one must be aware of, depend partially on the response variable. If first arrival dates are used, the problems are those of representative sampling. It is not always certain that sampling has been similar during the whole study area, particularly so in "bird club"-type of data. Possible bias in sampling should be checked and controlled for, if significant. If mean migration time is the response variable (or any point estimate of the whole migration) the sampling problems are less likely, but there are also problems arising from the target species' populations. In some areas, breeding populations are mixing with birds passing by. There is no easy way, perhaps no way at all, to separate migrants from breeders at least in the data collected in the past. Therefore, studies made on sites, such as remote bird station islands, have some benefit, although remoteness may cause other factors that increase error variance. Another strict statistical problem is that study seasons are often truncated from one or both ends. We are aware of one study where truncation has been accounted for by modelling the migration numbers appropriately (Jenni and Kéry, 2003).

If the data permit, the results obtained with these methods are straightforward to interpret. The variables and dataset used may contain several properties, which appreciably reduce the validity of these methods. It is likely that successive years' values are not independent and the change with year is often not linear. While we are not particularly worried about the independence of yearly observations, and in our experience, autocorrelation is not a major issue; non-linearity of timing changes is a real complication. In particular, it complicates comparisons between published results, because authors have used variable periods without noticing the relevant and underlying pattern of climate change. The recent climate warming period is said to have started in the 1960s (Walther *et al.*, 2002), but also later "starts" have been defined, e.g., late 1970s (Klein Tank and Können, 2003) and late 1980s (Zalakevicius, 2001a,b), probably because of differences in change of regional and seasonal temperatures. In general, all too often regional and seasonal pattern of climate variation has been neglected. The analysed arrival datasets often start years before this last identified "change point" of climate trend. Some authors have split the time series in separate sets for "before" and "after" the start of warming to avoid the problem of non-linear change. We, however, do not recommend division of time series in subsets since, the results in papers that use different change points are difficult to compare.

The regression coefficients are sensitive to period length mainly because the longer the time series the more likely it is that the prevailing trend changes within the period, or there is long wave periodicity in the phenomenon studied (for long arrival time data series, e.g., Ahas, 1999; see also Figure 2). The effect of non-linearity should therefore be taken into account. One way to do this is to use curve fitting methods, or a smoothing method such as lowess (locally weighted scatterplot smoother, Cleveland, 1981), but often statistical tests are unavailable for the latter. In the special case of recent climate change, it would be the best first solution to analyse only data that covers the warming period, i.e., from *c.* 1970 onwards.

There have been only a few attempts in which the factors explaining differences in species-specific responses have been modelled with higher-class statistical models. The most elaborate models used are those of Forchammer *et al.* (2002) and Jenni and Kéry (2003). The former analysed the effect of NAO with autoregressive phenological models and the latter used iterative least squares methods to achieve maximum likelihood estimates for the random and fixed model factors. While the change in timing can be proven with rather elementary statistical methods, it is crucial to use advanced statistical modelling to reveal how the effect of various factors can explain the variance in changes of timing.

VII. CONFOUNDING FACTORS

Simple analyses of temperature effect and trend are inevitably coarse. Arrival and departure dates depend on other factors as well. Some of these have been identified and incorporated in the models. We divide the confounding factors in two groups: (1) those depending on the observers, and (2) those depending on birds. An example of the former is the possible trend of observer activity and skill (Sparks *et al.*, 2001; Butler, 2003; Vähätalo *et al.*, 2004, but *cf.* Mason, 1995; Tryjanowski *et al.*, 2002), and of the latter, the trend in population size (Sparks, 1999; Huin and Sparks, 2000; Sparks *et al.*, 2001; Tryjanowski and Sparks, 2001). If any such confounding variable, that is not included in the model, correlates with year and/or temperature, correlation and regression coefficients or their standard errors are likely to be affected. These factors could be partialled out in correlation or added in regression models. This may be a bit controversial in some cases, however, because straightforward removal of trend by including year in the model is likely to be incorrect if the "cause" of the year effect is a weather factor varying collinearly with it. There is clearly a lot to be done to achieve higher accuracy of trend and temperature effects.

VIII. OVERVIEW OF CHANGES OF TIMING IN THE LATTER PART OF THE 20TH CENTURY

Burton (1995), Zalakevicius and Zalakeviciute (2001), Bairlein and Winkel (2001) and Fiedler (2001) have earlier summarised evidence for changes in arrival and departure dates of migratory birds. Impacts on birds have been included also in more general reviews and meta-analyses (Hughes, 2000; Root *et al.*, 2003). We give below an overview on the present state of affairs knowing that much work is currently under way and new reviews will soon be needed. We collected published time series of migration times that satisfy the following requirements: (1) they included the last warming period and did not start prior to 1940 and ended in the 1990s, (2) they covered at least 15 years, and (3) trends of change in arrival/departure and/or effect of temperature were expressed as days/year or days/°C or could be recalculated as such. We relaxed the last requirement in a few cases when compiling summaries of direction of effect and significances of them, if this information could be extracted from the paper. We could not take into account possible effects of confounding factors listed above. Therefore, the degree of bias in average response calculations remains to be studied later. For example, if observer skills and efficiency have improved along with climate change, calculated trends are too strong. We like to add that most of the cited analyses included some comment that appeared to regard their data unbiased, and in a few datasets, we know that constant effort routines were applied. This review concentrates on Palaearctic data, because most analyses were from there. During the work, analyses from other areas, particularly from North America, started to be published and we attempted to include them at the last moment.

A. Arrival—Trend of Timing

1. Variation among species

The overall tendency in many areas is that arrival has become earlier in the majority of species (Table 2). The distribution of calculated regression coefficients is highly significantly skewed towards negative slopes both in most separate datasets and overall. In studies, where first arrival was the response variable 39% of 983 time series showed significantly negative coefficients against the less than 2% significantly positive ones. In the 222 datasets using mean migration time 26% of time series indicated significantly earlier and 5% significantly later arrival. Variation among species and samples is considerable, and deserves further study. While publication bias is always possible in new research directions, we feel that in this case it cannot be substantially involved.

The majority of species has thus shown earlier arrival towards the end of the 20th century in practically every larger recent dataset. Average change of first arrival date over all species and sites, −0.373 days/year, was highly significantly negative, i.e., the average species in widely separate areas in Eurasia tended, in recent years, to arrive earlier in the study areas at or near the breeding grounds (Table 3). The results for mean migration time are in the same direction but the strength of response is clearly smaller, −0.100 days/year, but statistically significant (Table 3). The difference between the two is partially due to the fact that mean migration data is based on better standardised datasets. It may also be that earliest migrants respond more strongly to climate change than the rest of the population (Sparks et al., in press; Tøttrup, 2004; Vähätalo et al., 2004). Butler's (2003) results from NE USA are, in general, suggesting similar changes as our Eurasian datasets.

When regression coefficients are plotted against the average migration time of each species, interesting patterns emerge as in the Lithuanian example (Figure 3, data from Zalakevicius and Zalakeviciute, 2001). Most species respond to recent climate warming, also the later migrating long-distance migrants. This was observed also by Loxton et al. (1998), Sokolov et al. (1998), Loxton and Sparks (1999), and others. There is a clear tendency towards smaller regression coefficients in later migrants. Later migrants are usually also long-distance migrants. Two explanations have been put forward to explain this pattern: (1) migration of long-distance species is under stronger endogenous control (Huin and Sparks, 2000; leaning on Gwinner (1986) and Berthold (1996)), and (2) variability of weather conditions is less in late than in early spring. This pattern is what most students of the subject would have expected on climate change patterns. If climate is warming in Europe more than in more southerly areas, then short-distance migrants are affected by these changes all year round, but

Table 2 Distribution of trends of arrival times in 26 different analyses

Site	Sources of data	Country	Location	Period	Type of location	Significantly negative	Negative (ns)	ns	Positive (ns)	Significantly positive	Number of species
FA											
Troms	Barrett (2002)	Norway	69°40N, 18°50E	1970–2000	R	1		12		1	14
Ob	Paskhalny (2002)	Russia	66°00N, 67°00E	1970–2001	R	8		1	3		12
Kemi-Tornio	Rainio and Lehikoinen (unpubl.)	Finland	65°50N, 24°20E	1967–2001	R	54	50	–	20	2	126
Turku	Rainio and Lehikoinen (unpubl.)	Finland	60°27N, 22°15E	1965–2001	R	94	57	–	10	1	162
Jurmo	Rainio and Lehikoinen (unpubl.)	Finland	59°50N, 21°37E	1970–2001	S	54	72	–	5	0	131
Estonia	Sokolov et al., (1998)	Estonia	58°50N, 25°40E	1969–1991	R	1	7				8
Aboyne	Jenkins and Watson (2000)	Great Britain	57°05N, 02°46W	1974–1999	R	5	12		6	1	24
Christiansø	Tøttrup (2004)	Denmark	55°19N, 15°11E	1976–1997	S	8	17	–	0	0	25
Vilnius	Zalakevicius (unpubl.)	Lithuania	54°40N, 25°16E	1971–2000	R	20	18		5	1	44
Zuvintas	Zalakevicius (2001a,b)	Lithuania	54°28N, 23°32E	1966–1995	R	50		1		5	56
Calf of Man	Loxton and Sparks (1999)	Great Britain	54°03N, 04°49W	1959–1998	S	4	10	–	6	0	20
Barguzinsky	Ananin (2002)	Russia	54°00N, 11°00E	1984–2001	R	34	70		20		124

(*Continued*)

15

Table 2 Continued

Site	Sources of data	Country	Location	Period	Type of location	Significantly negative	Negative (ns)	ns	Positive (ns)	Significantly positive	Number of species
Bardsey	Loxton and Sparks (1999)	Great Britain	52°45N, 04°48W	1959–1998	S	3	7	–	10	1	21
Wielko-polska	Tryjanowski et al. (2002)	Poland	52°17N, 16°43E	1970–1996	R	4	10		2	0	16
Skokholm	Loxton and Sparks (1999)	Great Britain	51°40N, 05°20W	1959–1998	S	0	9	–	8	2	19
Essex	Sparks and Mason (2001)	Great Britain	51°00N, 00°00E	1950–1998	R	9	13	–	9	1	32
l'Oise	Sueur and Triplet (2001)	France	51°00N, 04°00E	1970–2000	R	3	17	–	12	1	33
Somme	Sueur and Triplet (2001)	France	51°00N, 02°00E	1970–2000	R	4	9		18	4	35
Portland	Loxton and Sparks (1999)	Great Britain	50°30N, 02°30W	1959–1998	S	9	10	–	5	0	24
Leicester-shire	Loxton et al. (1998)	Great Britain		1966–1996	R	8		15			23
Sussex	Loxton, et al. (1998)	Great Britain		1966–1996	R	14	18	–	2	0	34
FA, sum						387		576		20	983
%						39%		59%		2%	

16

MMT

Jurmo	Rainio and Lehikoinen (unpubl.)	Finland	59°50N, 21°37E	1970–2001	S	28	66	–	27	10	131
Rybachy	Sokolov et al. (1998)	Russia	55°05N, 20°44E	1973–1990	S	11	14	–	1	1	27
Christianso	Tøttrup et al. (unpubl.)	Denmark	55°19N, 15°11E	1976–1997	S	1	19	–	5	0	25
Helgoland	Hüppop and Hüppop (2003)	Germany	54°11N, 7 °55E	1960–2000	S	17	6	–	1	0	24
Helgoland	Bairlein and Winkel (2001)	Germany	55°11N, 8 °55E	1961–1993	S	1	9	–	5	0	15
MMT, sum						58	114		39	11	222
%						26%	51%	0%	18%	5%	

Datasets are arranged in the table roughly from NE to SW. FA, first arrival dates; MMT, mean/median migration time; R, regional data; S, bird station data. See text for dataset selection details.

17

Table 3 Strength of response to year, North Atlantic Oscillation and local temperature in the time series concerning arrival of various migratory bird species in the latter half of the 20th century

Response type (unit)	Number of time series	Upper 95% confidence limit	Average response	Lower 95% confidence limit
Trend, FA (days/year)	590	− 0.342	− 0.373	− 0.403
NAO, FA (days/unit change of NAO-index)	128	− 3.382	− 3.959	− 4.535
Local temperature, FA (days/°C)	203	− 2.472	− 2.901	− 3.331
Trend, MMT (days/year)	225	− 0.137	− 0.100	− 0.223
NAO, MMT (days/unit change of NAO-index)	149	− 1.350	− 1.636	− 1.921
Local temperature, MMT (days/°C)	153	− 1.433	− 1.761	− 2.089

FA, first arrival; MMT, mean/median migration time of a population. See text for data selection. Note: time series were the same as listed in Table 2.

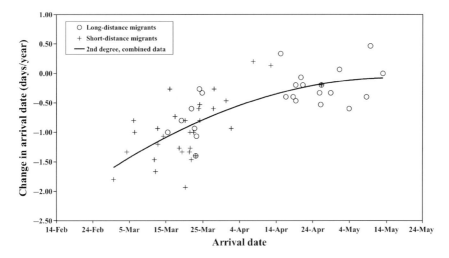

Figure 3 Change rates of first arrival dates (days/year) of 56 bird species in Lithuania (Žuvintas Strict Nature Reserve) in 1966–1995 in relation to average arrival day. Change rates calculated from the data in Žalakevičius (2001a). Regression (Day 1 = 1 January): Change $= -0.0003 \times \mathrm{Day}^2 + 0.071 \times \mathrm{Day} - 4.933$, $R^2 = 55.2\%$. Negative rate indicates earlier arrival.

the tropical and southern hemisphere Palaearctic migrants only in the latter part of their migratory journey.

Ecological traits, e.g., habitat preference, feeding ecology, etc., may also affect the type of response of individual species to climate change. In boreal and arctic zones, most waterfowl species are short-distance migrants moving only to areas where they avoid winter ice, while in other groups many species are long-distance migrants. In Finland and especially in the Arctic regions, the period of ice cover in lakes and in the Baltic Sea bays is a strong factor regulating start of breeding in Eider and other waterfowl, although resources with which to start breeding are under the influence of changes taking place in wintering and migration regions (Hario and Selin, 1988).

Loxton *et al.* (1998) and Loxton and Sparks (1999) stated that small insectivores showed the strongest tendency to earlier arrival and temperature sensitivity, but their studies did not include a wide spectrum of different feeding types and habitats, e.g., intra-European migrants, waterfowl and shorebirds were not well represented. According to Butler (2003) seed-eating grassland species of NE America advanced their spring arrival significantly more than other feeding ecology-habitat groups while response of scrub, aquatic and aerial birds did not differ from these and each other. None of these studies took into account the effect of phylogenetic correlations with habitat and feeding ecology classifications. We clearly need more thorough analysis of habitat and feeding ecology effects.

2. Variation between Areas

Response to climate change varies between areas (Sparks et al., 1999). Differences in response arise more probably from spatio-temporal variation of climate change than from differences in birds' behaviour. We still do not have enough published analyses, in which comparable methods and response measurements have been used, to perform continent-wide analyses. We therefore only present examples from Eurasia from Great Britain to eastern Siberia and from central to northern Europe without an in-depth analysis (Table 2). Contrary to most other parts of Europe, Peñuelas et al. (2002) observed a significant delay of 7–30 days in the arrival date of five from six sub-Saharan migrants to NW-Spain from 1952 to 2000, but population changes were not accounted for in the analysis. North American birds show similar responses to climate change in migration times as those in Eurasia (Bradley et al., 1999; Inouye et al., 2000). Butler (2003) analysed a dataset of 103 species from New York and Massachusetts and concluded that short-distance migrants advanced spring arrival by 13 days (0.28 days/year, calculated from the median years of Butler's periods, which do not fit well with the recent warming period) and long-distance ones by 4 days (0.09 days/year) from 1903–1950 to 1951–1993.

3. Spatial and Intraspecific Constancy of Trends

In our database on arrival dates, 22 species were studied in 10 or more areas. The set of species is not a random sample of all bird species and study areas are concentrated in Europe, but it offers a possibility to look at the constancy of trend within species (Table 4). In six species from sand martin Riparia riparia to pied flycatcher Ficedula hypoleuca, separate trend estimates are skewed statistically significantly to negative values. If we look only at significantly negative trends partially the same species emerge, but also species lower in the list have rather high proportions of time series suggesting earlier arrival in recent years. This may be, because in some species only some populations are arriving earlier. Further analysis of spatial variation of trends is clearly a worthy effort.

B. Arrival—Relationship to Temperature and Weather Systems

Local temperature at the site of observation has been used as the explanatory variable in most studies. Authors can collect relatively easily their local temperatures whereas migration route temperatures have been more difficult to

Table 4 Distribution of trends of change in arrival dates in the species, which have been studied most frequently

	fa	Mean	Median	Total	Neg s	Neg	ns	Pos	Pos s	Proportion of significantly negative changes (%)	Proportion of negative changes (%)
Sand martin	12		1	13	7	6	0	0	0	53.8	100.0
Blackcap	9	2	2	13	8	4	0	1	0	61.5	92.3
Chiffchaff	9	2	2	13	7	4	0	2	0	53.8	84.6
White/pied wagtail	7	3	1	11	2	7	1	1	0	18.2	81.8
Barn swallow	13	1	1	15	5	7	1	2	0	33.3	80.0
Pied flycatcher	10	2	3	15	5	7	0	3	0	33.3	80.0
Sedge warbler	11	1		12	2	7	2	1	0	16.7	75.0
Tree pipit	12	1	2	15	6	5	3	0	1	40.0	73.3
House martin	13		1	14	4	6	1	3	0	28.6	71.4
Willow warbler	15	3	2	20	10	4	2	3	1	50.0	70.0
Yellow wagtail	12		1	13	1	8	2	1	1	7.7	69.2
Redstart	12	2	1	15	3	7	2	3	0	20.0	66.7
Wheatear	10		1	11	4	3	1	2	1	36.4	63.6
Garden warbler	12	2	2	16	4	6	0	4	2	25.0	62.5
Wood warbler	8	1	2	11	4	2	2	3	0	36.4	54.5
Lesser whitethroat	9	2	2	13	4	3	3	3	0	30.8	53.8
Common swift	13	1	1	15	2	6	3	3	1	13.3	53.3

(Continued)

Table 4 Continued

	fa	Mean	Median	Total	Neg s	Neg	ns	Pos	Pos s	Proportion of significantly negative changes (%)	Proportion of negative changes (%)
Whitethroat	11	2	2	15	2	6	2	4	1	13.3	53.3
Common sandpiper	8	1	1	10	2	3	0	5	0	20.0	50.0
Spotted flycatcher	12	1	1	14	1	6	5	3	0	6.7	46.7
Whinchat	13	1	2	15	3	3	1	7	1	20.0	40.0
Cuckoo	14	1	2	17	2	4	3	4	3	12.5	37.5
Proportion					29%	37%	11%	19%	4%		
Number					88	114	34	58	12		

Tendency to earlier arrival is indicated by high proportion of negative trends. Explanations: neg s = significantly (at $p < 0.05$) negative trend, neg = negative non-significant trends, ns = non-significant trend, direction not given in the original source, pos = positive non-significant trend, and pos s = significant (at $p < 0.05$) positive trend. Most time series are for first arrival (fa), a few for mean or median arrival of whole "populations". Note: time series were the same as listed in Table 2.

22

obtain in the past. We believe that all authors have realised that birds cannot directly predict the conditions in the target area of their next flight. The authors use temperature of the study area just as an indicator of weather in a larger area. From meteorological analyses, we know that this is justified for fairly large areas (Heino, 1994; Sokolov *et al.*, 1998). In addition to temperature, many other weather parameters may play a role (Alerstam, 1990; Zalakevicius, 1993; Elkins, 1995; Zalakevicius *et al.*, 1995), but distinction between their independent effects may prove difficult, e.g., proportion of tail winds correlates positively with temperature (Sparks *et al.*, 2001). In a few recent works, the analyses have broadened to more general indicators. Huin and Sparks (1998) explained the arrival of barn swallows *Hirundo rustica* in Britain by temperatures south of Britain in France and Iberia and Hüppop and Hüppop (2003) and Vähätalo *et al.* (2004) looked at the North Atlantic Oscillation (NAO) as a general mechanism affecting climate in Europe. Analyses, which use weather along the migration route as independent variables, are the next step, which is already taking place (Ahola et al., 2004). These analyses will not be as straightforward to use as might appear at first sight. For example, a careful spatio-temporal analysis of the progress of migration must accompany the analysis of weather impacts.

Temperature and climate analyses have proven that birds in general respond to high temperature and positive NAO phases by arriving earlier. We summarise patterns of several regression and correlation analyses in Table 3. Sparks *et al.* (2002) suggested that the typical strength of temperature response in bird arrival was 2 days/°C, which was less than the 6–8 days/°C response in plants. In the datasets collected for this review the response varies from 2.47 to 3.33 days/°C ($n = 203$ regressions from Eurasia), not much different from their figure. Despite this, Root *et al.* (2003) did not find differences in trends between birds, invertebrates, amphibians and non-trees (earlier by 5–6 days/decade), but trees (3 days/decade) showed smaller response than others. The response variables for birds were, however, a mixture of migration and breeding responses. Large-scale comparison can tell little about possible timing problems, since responses at community scale can still be different (Harrington *et al.*, 1999; Both and Visser, 2001).

C. Departure

Responses determining departure date probably depend on the individual species. For some species, advancement in spring may result directly in earlier departure in autumn while in others longer summer and later autumn may lead to later departure. Apparently, the direction of response depends on the importance of summer and winter seasons in the species' complete life history strategy.

Amateur ornithologists continually register records of delayed autumn migration or peculiar first winter observations for different areas, which strongly suggest that delaying migration may be a climate response. Rigid analyses are rare (Gatter, 1992; Bairlein and Winkel, 2001; Fiedler *et al.*, 2004, this volume). Gilyazov and Sparks (2002) reported departure of seven species from the Kola peninsula from 1931 to 1999 by calculating average departure times for three time windows 1931–1941, 1958–1978 and 1979–1999 and reported trends for the whole period. If we only look at the two latter periods, the change in departure date seems to be rather small. Pine grosbeak *Pinicola enucleator* (rate of change: − 0.36 days/year) and brambling *Fringilla montifringilla* (−0.45 days/year) departed earlier and the other five species departed later (0.18–0.64 days/year) in 1979–1999 than in 1958–1978. Waterfowl, the whooper swan *Cygnus cygnus* and goldeneye *Bucephala clangula* departed significantly later in warmer autumns. Delay of autumn migration of waterfowl is a well-known phenomenon around the Baltic as well (Hario *et al.*, 1993). Early autumn migrants in the Kola Peninsula departed earlier in warm autumns (August temperature used, wood sandpiper *Tringa glareola*, $r = -0.53$, Arctic tern *Sterna paradisaea* −0.47, osprey *Pandion haliaetus* −0.29, willow warbler *Phylloscopus trochilus* −0.22).

Sparks and Mason (2001) studied the arrival and departure dates as well as length of stay (duration of summer residence) in Essex, England. The distribution of departure changes is significantly skewed to the positive side, i.e., departure is becoming later in more species than expected by chance. In eight species, delaying autumn migration was statistically significant, the delay varying between 0.3 and 1.4 days/year from the redstart *Phoenicurus phoenicurus* to the hobby *Falco subbuteo*.

All above studies concerned the last observations in a study area. Little is so far known about the population averages. Sokolov *et al.* (1998) analysed Rybachy's long-term data for autumn departure mean migration dates and their dependence on temperature of preceding months (Figure 4). In their data, spring arrival is directly reflected in the autumn departure date of 14 species, i.e., the warmer the spring was the earlier the species arrived and the earlier they departed. The species studied were mostly small insectivorous passerines, mostly with one annual brood. There was no overall difference in the response to spring earliness between long- and short-distance migrants. Timing of juvenile dispersal correlated positively with autumn departure in all 13 examined species and significantly so (at $p < 0.05$) in five of them with no evident difference between short- and long-distance migrants. In two multi-brooded species, the great tit *Parus major* and the chaffinch *F. coelebs*, late breeding led to delayed autumn departure.

In Southern Germany median autumn migration was delayed by on average 3.4 days in 19 out of 28 species studied in 20 years from 1970 (Gatter, 1992; Bairlein and Winkel, 2001). On Helgoland, one of seven

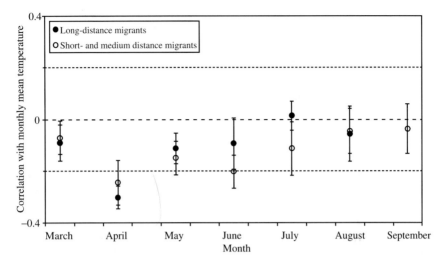

Figure 4 Dependence of autumn departure on average monthly temperatures of April through September in Rybachy. Data from Sokolov *et al.* (1998) averaged over migrant species grouped according to distance of migration. Confidence limits (95%) are shown with vertical bars. *Species studied: Long-distance migrants*: icterine warbler, wood warbler, red-backed shrike, tree pipit, lesser whitethroat, whitethroat, white wagtail, pied flycatcher, garden warbler, willow warbler, spotted flycatcher, redstart. *Medium- and short-distance migrants*: blackcap, robin, chiffchaff, dunnock, chaffinch, blue tit, meadow pipit, reed bunting, great tit, brambling, goldcrest.

short-distance migrants (wren *Troglodytes troglodytes*) and three of nine long-distance migrants (willow warbler, redstart, and ring ouzel *Turdus torquatus*) delayed significantly their autumn migration. The net result in those species that were studied in both seasons was increasing the time span between average spring and autumn migrations in 11 of 15 species, the swift *Apus apus* providing the opposite extreme with reducing time span (Bairlein and Winkel, 2001).

Jenni and Kéry (2003) analysed Swiss data of passing autumn migrants. Their main observation was advancement of autumn migration in long-distance tropical migrants and delay in short-distance migrants, a result in concordance with Rybachy's results. Of the 25 long-distance migrants 20 advanced their autumn migration while only 12 out of 40 short-distance species did the same (all species classified whether the change was statistically significant or not). The authors suggested this was due to selection pressure caused by the start of the Sahelian dry period in autumn. The number of broods had a significant effect on the advancement of autumn departure—long-distance single-brooded species advanced their migration time while species with one or two broods did not. In short-distance migrants the relationship to the number of broods was even more complicated: the species with variable number of broods delayed autumn migration more than single-brooded and double-brooded species. What might have been the role of advanced spring migration (Sokolov *et al.*, 1998, above) in the Swiss data remained unexplored.

D. Length of Stay

Length of stay in the breeding grounds is determined ultimately by the time needed for breeding and, in many species, for performing the annual moult. If the length of summer is increasing at either end, it may have consequences for the timing of birds' annual cycle. Three alternatives are possible for birds to respond to the lengthening of summer: (1) length of stay and durations of breeding and moult remain constant; (2) length of stay does not change, but timing of events (arrival, breeding, moult, and departure) shifts earlier; (3) length of stay increases with accompanying changes in arrival and/or departure. So far, it is not at all clear how different species respond. Predictions can be derived based on either the benefits of a longer stay during the breeding and moult periods or a shorter stay because of factors, which favour earlier/faster migration or earlier arrival at moulting or wintering areas. Given the diversity of species-specific annual strategies, generalisations that are valid for most bird species are unlikely to be found.

The easiest data to look at for investigating this question is first arrival–last departure observations. It is of little use for evaluating the biological significance of lengthening of the summer residence of birds, because it is not individual information and does not well represent population averages. The next best choice is the data from bird stations, which operate both in spring and in autumn. In bird station data, the different timing of young and adult birds can be taken into account, if ringing data is used. So far, few such analyses have been made. The shortcomings in these data are (1) that mean migration times in spring and autumn in a given station are not necessarily connected with the same breeding populations, and (2) they do not relate to individual birds. The first problem is smaller, but exists since populations are mixed in *en route* migration data. The second problem can hardly be overcome unless individually marked birds could be monitored year round as in satellite tracking.

In the Kola peninsula (Gilyazov and Sparks, 2002) the length of stay increased significantly in four of seven species studied, on average by 12 days, while it shortened by 16 days in the whooper swan. This order of lengthening may give a better chance, e.g., to lay a repeat clutch, but it hardly can increase the frequency of second clutches in any species. On Helgoland (Bairlein and Winkel, 2001) seven short-distance migrants lengthened their time span significantly by nine days and nine long-distance migrants only by less than one day (not significant, but if a possible outlier, the swift, is omitted the lengthening is four days, and there is no effect of migration distance). In Essex, England (Sparks and Mason, 2001) the length of stay increased significantly in eight species and the distribution for 32 species' regression coefficients was significantly skewed towards lengthening of stay.

IX. CONSEQUENCE OF CHANGING TIMING

Large changes in arrival and departure dates may have important consequences for other parts of a species' annual cycle. The final step—a species or an appreciable part of its population becoming sedentary—has also been documented with different example species in different areas (Zalakevicius et al., 1995; Berthold, 1998; Fiedler et al., 2004, this volume). Partial migrants are continuously "monitoring" the weather and food conditions and they can be especially fast in taking whatever advantage might be available, exemplified by, e.g., the greenfinch *Carduelis chloris*, which has increased its winter population in Finland 6-fold in the last 40 years (Väisänen and Solonen, 1997) and become simultaneously a rarer winter visitor to Northern Germany. Many Lithuanian species have shortened their migration distance (Svazas et al., 2001).

Correlation of arrival and breeding dates is significant in many, though not all, species and geographical areas (Sokolov et al., 1998; Sparks et al., 2001; Sippola, 2003). It is, however, not straightforward, to predict the fitness consequences of possibly relaxing time constraints for breeding and moult. In some species and in some areas (e.g., summer visitors in northernmost Europe) earlier and longer summer may be beneficial, but in others, accompanying changes in food sources may lead to other constraints. A few recent studies have concentrated on the mismatch-problems arising from these changes (Both and Visser, 2001). To understand thoroughly the fitness consequences of climate induced changes in timing of the annual schedule of birds, we need basic research on weather dependence of breeding success (Eeva et al., 2002), which has remained in the background in recent decades. These questions are looked at in more detail in other chapters of this book.

The community level prediction put forward by Berthold (1991), and expressed later in different forms (Zalakevicius, 2001a,b; Jenni and Kéry, 2003), that the bird community compositions may change, if climate warming favours sedentary and short-distance migrant species more than long-distance migrants, cannot be tested by studying changes of migration times. Their documentation is, however, an important first step in disentangling both individual fitness, population and community consequences.

X. FUTURE WORK

The quality of data still needs further study. There are several points of concern as pointed out in this review, but despite this, the sign of change and the signal are most likely correct. What needs to be done is to identify and quantify the other factors that cause arrival and departure dates to shift in the same direction as climate and weather. At least in some areas, detectability of first arrivals may have changed and there is no careful analysis on how important a factor this might be. Standardised observation has been the routine in many bird stations and therefore these data might be superior to the first arrival data of local bird clubs and similar organisations. Most progress could be achieved if the available good quality data of migrants could be collected in a joint database, which offered a variety of possible research directions.

The approaches used in the papers published so far have been a mixture of what can be used. The diverse ways the results are summarised in articles is an obstacle for a meta-analysis. We used here summaries of simple linear regressions and correlations to show the trends and temperature effects, mainly because these have been the most common methods used so far. We also had to accept that time series lengths differed between published studies. It was not possible to recalculate the effects for periods that are more similar.

In the near future, we expect that the patterns that we showed to exist over bird species in general and in large geographical areas will be better understood. The understanding of ecological traits of species, their responsiveness to climate change and the consequences to the whole life history of different species should be the next step on which to concentrate.

REFERENCES

Ahas, R. (1999) *Int. J. Biometeorol.* **42**, 119–123.
Ahola, M., Laaksonen, T., Sippola, K., Eeva, T., Rainio, K. and Lehikoinen E. (2004) Variation in climate warming along the migration route uncouples arrival and breeding dates. *Global Change Biol.* **10**, 1–8(doi: 10.1111).
Åkesson, S. and Hedenström, A. (2000) *Behav. Ecol. Sociobiol.* **47**, 140–144.
Alerstam, T. (1990) *Bird Migration.* Cambridge University Press, Cambridge.
Ananin, A.A. (2002) In: *Long-term Dynamic of Bird and Mammal Populations and Global Climatic Changes* (Ed. by O.V. Askeyev), pp. 107–112. Institute of Natural Systems Ecology, Kazan.
Bairlein, F. and Winkel, W. (2001) In: *Climate of the 21st Century: Changes and Risks* (Ed. by J.L. Lozan, H. Grassl and P. Hupfer), GEO, Hamburg.
Barrett, R.T. (2002) *Bird Study* **49**, 270–277.
Berthold, P. (1991) *Acta Congr. Int. Ornithol.* **20**, 780–786.

Berthold, P. (1996) *Control of Bird Migration*. Chapman & Hall, London.

Berthold, P. (1998) *Naturwissenschaftliche Rundschau* **51**, 337–346.

Both, C. and Visser, M. (2001) *Nature* **411**, 296–298.

Bradley, N.L., Leopold, C.A., Ross, J. and Huffaker, W. (1999) *Proc. Natl Acad. Sci. USA* **96**, 9701–9704.

Burton, J. (1995) *Birds and Climate Change*. Helm, London.

Butler, C.J. (2003) *Ibis* **145**, 484–495.

Cleveland, W.S. (1981) *Am. Stat.* **35**, 54.

Coppack, T. and Both, C. (2002) *Ardea* **90(3)**, 369–378.

Eeva, T., Lehikoinen, E., Lummaa, V., Rönkä, M. and Currie, D. (2002) *Ecography* **25**, 705–713.

Elkins, N. (1995) *Weather and Bird Behaviour*, 2nd Ed., Poyser, London.

Fiedler, W. (2001) In: *Avian Migration* (Ed. by P. Berthold, E. Gwinner and E. Sonnenschein), pp. 21–38. Springer, Berlin.

Forchammer, M.C., Post, E. and Stenseth, N.C. (2002) *J. Anim. Ecol.* **71**, 1002–1014.

Frederick II of Hohenstaufen (1244–48) *De Arte venandi cum avibus* (translated by C.A. Wood and F.M. Fyfe (1943). The art of falconry: being the De Arte venandi cum avibus of Frederick II of Hohenstaufen. Stanford University Press, Stanford).

Gatter, W. (1992) *J. Orn.* **133**, 427–436.

Gilyazov, A. and Sparks, T. (2002) *Avian Ecol. Behav.* **8**, 35–47.

Gwinner, E. (1986) *Circannual Rhythms: Endogenous Annual Clocks in the Organization of Seasonal Processes*. Springer, Berlin.

Gwinner, E. (1996) *J. Exp. Biol.* **199**, 39–48.

Hario, M. and Selin, K. (1988) *Finnish Game Res.* **45**, 3–10.

Hario, M., Lammi, E., Mikkola, M. and Södersved, J. (1993) *Suomen Riista* **39**, 21–32.

Harrington, R., Woiwod, I.P. and Sparks, T.H. (1999) *Tree* **14**, 146–150.

Heino, R. (1994) *Finn. Meteorol. Inst. Contrib.* **12**, 1–209.

Helbig, A. (2001) In: *Avian Migration* (Ed. by P. Berthold, E. Gwinner and E. Sonnenschein), pp. 3–20. Springer, Berlin.

Hughes, L. (2000) *Tree* **15**, 56–61.

Huin, N. and Sparks, T.H. (1998) *Bird Study* **45**, 361–370.

Huin, N. and Sparks, T.H. (2000) *Bird Study* **47**, 22–31.

Hüppop, O. and Hüppop, K. (2003) *Proc. R. Soc. Lond., Ser. B Biol. Sci.* **270**, 233–240.

Inouye, D.W., Barr, B., Armitage, K.B. and Inouye, B.D. (2000) *Proc. Natl Acad. Sci. USA* **97**, 1630–1633.

Jenkins, D. and Watson, A. (2000) *Bird Study* **47**, 249–251.

Jenni, L. and Kéry, M. (2003) *Proc. R. Soc. Lond. B* **270**, 1467–1471.

Justander, J.G. (1786) Specimen calendarii florae et faunae aboensis. *Aboœe*.

Kington, J. (Ed.) (1988) *The Weather Journals of a Rutland Squire. Thomas Barker of Lyndon Hall*. Rutland County Museum, Oakham.

Klein Tank, A.M.G. and Können, G.P. (2003) *J. Clim.* **16**, 3665–3680.

Kok, O.B., Van Ee, C.A. and Nel, D.G. (1991) *Ardea* **79**, 63–65.

Krementz, D.G., Seginak, J.T. and Pendleton, G.W. (1994) *Wilson Bull.* **106**, 482–493.

Leche, J. (1763) Kongl. *Svenska Vetenskaps Akademiens Handlingar* **24**, 257–268.

Liechti, F. and Bruderer, B. (1998) *J. Avian Biol.* **29**, 561–568.

Linné C. (1757) Calendarium florae eller Blomster-Almanack. Stockholm. Reprint, 1977.

Loxton, R.G. and Sparks, T.H. (1999) *Bardsey Observ. Rep.* **42**, 105–143.

Loxton, R.G., Sparks, T.H. and Newnham, J.A. (1998) *Sussex Bird Rep.* **50**, 182–196.

Mason, C.F. (1995) *Bird Study* **42**, 182–189.

Nervander, J.J. (1846) *Saima* **18**.

Paskhalny, S.P. (2002) In: *Long-term Dynamic of Bird and Mammal Populations and Global Climatic Changes* (Ed. by O.V. Askeyev), pp. 151–156. Institute of Natural Systems Ecology, Kazan.

Peñuelas, J., Filella, I. and Comas, P. (2002) *Global Change Biol.* **8**, 531–544.

Ptaszyk, J., Kosicki, J., Sparks, T.H. and Tryjanowski, P. (2003) *J. Orinthol.* **144**, 323–329.

Richardson, W.J. (1990) In: *Bird Migration: The Physiology and Ecophysiology* (Ed. by E. Gwinner), pp. 78–101. Springer, Berlin.

Root, T.L., Price, J.T., Hall, K.R., Schneider, S.H., Rosenzweig, C. and Pounds, J.A. (2003) *Nature* **421**, 57–60.

Sagarin, R. (2002) In: *Wildlife Responses to Climate Change: North American Case Studies* (Ed. by S.H. Schneider and T.L. Root), pp. 127–163. Island Press, Washington.

Sippola K. (2003) M.Sc. thesis, University of Turku (in Finnish).

Sokolov, L.V., Markovets, M. Yu., Shapoval, A.P. and Morozov, Yu.G. (1998) *Avian Ecol. Behav.* **1**, 1–21.

Sokolov, L.V., Markovets, M.Y. and Morozov, Y.G. (1999) *Avian Ecol. Behav.* **2**, 1–18.

Sparks, T.H. (1999) *Int. J. Biometeorol.* **42**, 134–138.

Sparks, T.H., Bairlein, F., Bojarinova, J.G., Hüppop, O., Lehikoinen, E., Rainio, K., Sokolov, L.V., Walker, D. (in press) *Global Change Biology*.

Sparks, T.H. and Carey, P.D. (1995) *J. Ecol.* **83**, 321–329.

Sparks, T.H. and Braslavská, O. (2001) *Int. J. Biometeorol* **45**, 212–216.

Sparks, T.H. and Mason, C.F. (2001) *Essex Bird Rep.* **1999**, 154–164.

Sparks, T.H., Heyen, H., Braslavska, O. and Lehikoinen, E. (1999) *BTO News* **223**, 8–9.

Sparks, T.H., Roberts, D.R. and Crick, H.Q.P. (2001) *Avian Ecol. Behav.* **7**, 75–85.

Sparks, T., Crick, H., Elkins, N., Moss, R., Moss, S. and Mylne, K. (2002) *Weather* **57**, 399–410.

Sueur, F. and Triplet, P. (2001) *Avifaune picarde* **1**, 111–120.

Svazas, S., Meissner, W., Serebryakov, V., Kozulin, A. and Grishanov, G. (2001) *Changes of Wintering Sites of Waterfowl in Central and Eastern Europe*, 152 pp. Institute of Ecology, Vilnius.

Tryjanowski, P. and Sparks, T.H. (2001) *Int. J. Biometeorol.* **45**, 217–219.

Tryjanowski, P., Kuzniak, S. and Sparks, T. (2002) *Ibis* **144**, 62–68.

Tøttrup, A.P. (2004) M.Sc. thesis, Zoological Museum, University of Copenhagen.

Walther, G.R., Post, E., Convey, P., Merzel, A., Parmesan, C., Beebee, T.J.C., Fromentin, J.-M., Hoegh-Guldberg, O. and Bairlein, F. (2002) *Nature* **416**, 389–395.

Vähätalo, A., Rainio, K., Lehikoinen, A. and Lehikoinen, E. (2004) *J. Avian Biol.* **35**, 210–216.

Väisänen, R.A.V. and Solonen, T. (1997) *Linnut (The Yearbook of the Linnut-magazine)*, 70–97.

von Haartman, L. (1956) *Societas Scientiarum Fennica Årsbok—Vuosikirja* **33**(3), 1–23.

Zalakevicius, M. (1993) *Acta Orn. Lithuanica* **7–8**, 16–26.

Zalakevicius, M. (2001a) *Acta Orn. Lithuanica* **11**, 200–218.

Zalakevicius, M. (2001b) *Acta Orn. Lithuanica* **11**, 332–339.

Zalakevicius, M. and Zalakeviciute, M. (2001) *Folia Zool.* **50**(1), 1–17.

Zalakevicius, M., Svazas, S., Stanevicius, V. and Vaitkus, G. (1995) *Bird Migration and Wintering in Lithuania.* Acta Zoologica Lituanica: Ornithologia, Vol. 2, 252 pp. Institute of Ecology, Vilnius, a monograph (in English).

Migratory Fuelling and Global Climate Change

FRANZ BAIRLEIN* AND OMMO HÜPPOP

I. SUMMARY

Climate-induced changes on habitats are likely to have impacts on staging, stopover ecology and fuelling in migratory birds. The effects of these changes on migratory birds are very speculative due to the lack of detailed studies and the uncertainty in climate models with respect to geographical patterns of changes but pronounced regional and species-specific differences are likely. Terrestrial birds and those using inland wetlands are likely to face more pronounced environmental challenges during migration than coastal migrants. Staging migrants may suffer from deteriorating habitats but they may, on the other hand, be able to counteract adverse conditions owing to considerable plasticity in their migratory performance.

II. INTRODUCTION

Climate change affects ecosystems, habitats and species with increasing velocity and continuity (e.g., Lindbladh *et al.*, 2000; Walther *et al.*, 2002; Berry *et al.*, 2003; Parmesan and Yohe, 2003; Root *et al.*, 2003). Several

---

E-mail address: franz.bairlein@ifv.terramare.de (F. Bairlein)

ADVANCES IN ECOLOGICAL RESEARCH, VOL. 35
0065-2504/04 $35.00 DOI 10.1016/S0065-2504(04)35002-6

migratory birds react to increased local temperatures or to large-scale climatic phenomena such as the North Atlantic Oscillation (NAO) with changes in arrival and departure phenologies (Bairlein and Winkel, 2001; Buttler, 2003; Hüppop and Hüppop, 2003; Jenni and Kéry, 2003; Strode, 2003; Lehikoinen et al., chapter "Migratory Fuelling and Global Climate Change"). The NAO has a major influence on temperature and precipitation in Europe (Hurrell, 1995; Forchhammer et al., 2002). In years with a positive NAO-index, reflecting higher spring temperatures and higher precipitation in north-western Europe, birds arrive earlier than in other years. In barn swallows *Hirundo rustica*, spring arrival to a Danish breeding colony is related to the environmental conditions in Northern Africa, as shown by a strong positive correlation between annual mean arrival and the normalised difference vegetation index (NDVI) in spring in Algeria with mean arrival dates earlier in years with adverse environmental conditions in Northern Africa (A.P. Møller, manuscript). Both, NAO-Index and NDVI in Northern Africa show clear temporal trends with increasing numbers of winters with positive NAO-indices and deteriorating vegetation conditions during recent years (Hurrell, 1995; Forchhammer et al., 2002; A.P. Møller, manuscript).

These data suggest strong effects of climate change not only on local environmental conditions but also on staging and stopover ecology, which are worth to be considered in predicting impacts of climate change on migratory birds. However, general predictions are difficult as climate models predict pronounced differences in the change of surface air temperatures and precipitation on a regional scale and because of species-specific differences in migration "strategies", migration routes, speed, stopover and wintering areas.

However, seeking answers and predicting possible impacts on migratory performance must remain very cursory and very speculative as hardly any study has addressed these questions. Thus, we do not aim to provide an extended speculative approach rather to highlight certain aspects and scenarios derived from general knowledge of the control of bird migration (e.g., Gwinner, 1990; Berthold, 1996; Berthold et al., 2003) and of the temporal and spatial patterns of migration (e.g., Zink, 1973, 1975, 1981, 1985; Zink and Bairlein, 1995; Fransson and Pettersson, 2001; Wernham et al., 2002) and migratory fuelling (e.g., Bibby and Green, 1981; Bairlein, 1991a, 1998; Schaub and Jenni, 2000a, 2001a,b; Bairlein, 2003; Jenni-Eiermann and Jenni, 2003). Moreover, we concentrate on examples within the European–African bird migration system, because both widespread changes in migration phenology and scenarios for landscape changes are less demonstrated for any other of the global migration system, including the Nearctic–Neotropical system (Buttler, 2003). In addition, climate change appears to be much more pronounced in Central, Western and Northern Europe (http://www.ncdc.noaa.gov/img/climate/globalwarming/ipcc09.gif).

III. CHANGING STOPOVER HABITATS

The NDVI does not show a consistent temporal pattern on a regional scale (see web page at http://edcintl.cr.usgs.gov/adds/adds.html) nor do the effects of the NAO (e.g., Hurrell, 1995; Forchhammer et al., 2002). The latter show a dichotomous variation in influence on local weather across a meridional gradient in Europe. Higher winter temperature and precipitation are associated with high NAO-indices in Northern Europe but these conditions are associated with low NAO-indices in Southern Europe (Forchhammer et al., 2002). Moreover, there is a west–east gradient within Europe with more pronounced NAO effects in coastal North, Central and West Europe than in Eastern Europe (Ottersen et al., 2001; Visbeck et al., 2001).

NAO has not only consequences for regional climate in Europe. While north-western Europe experiences warmer and wetter winters during high NAO-indices, such winters are followed by decreased vegetation productivity in the African Sahel zone (Oba et al., 2001; Forchhammer et al., 2002; Wang, 2003), due to a close linkage between NAO and the African monsoon system (Zahn, 2003). For the Sahel, increasing desertification is predicted due to a southward shift of summer rain. The annual rate of desertification in the Sahel is about 0.5%, which corresponds to an area of about 80,000 km^2 affected by degradation (Pilardeaux and Schulz-Baldes, 2001). In Southern Africa, however, vegetation productivity increases during years with high NAO-indices (Oba et al., 2001).

Regional changes are also predicted for the Mediterranean Basin although the predicted scenarios are not yet clear in details. Several models predict a decrease of winter precipitation for the Iberian Peninsula as well as an increase in surface ambient temperature resulting in drier conditions and extended dry periods (Cubasch, 2001). In north-eastern Spain, cold–temperate ecosystems are progressively replaced by Mediterranean ones owing to progressively warmer conditions (Peñuelas and Boada, 2003).

In conclusion, current climate change thus is likely to increase areas considered to be turning into desert or drier habitats in Spain, Northern Africa and sub-Sahara, and this pattern is likely to intensify in the future. Migratory birds may need to cross larger distances of hostile habitat, and suitable staging grounds may become smaller and more dispersed. Consequently, habitat conditions for migrants will change considerably which will affect stopover performance of migrant bird species. Thus, questions arise whether migratory birds from Europe are able to accommodate to such changes, whether there are physiological limits to accommodation, and which species or fractions of populations of migratory birds will be able to cope best.

IV. MIGRATORY FUELLING AND SUCCESSFUL MIGRATION

Migration is energetically costly. Thus, many migratory species accumulate large amounts of energy reserves prior to migratory flights, of which most is fat (Biebach, 1996; Lindström, 2003). The mass of stored reserves may amount to more than 100% of lean body mass with maximum levels obtained by species crossing inhospitable areas such as sea and deserts with no feeding opportunities. The garden warbler *Sylvia borin*, for example, a long-distance European migratory songbird wintering in tropical Africa, weighs about 16–18 g during the breeding and wintering seasons, but increases its body mass to up to 37 g just before leaving to cross the Sahara, both in autumn and spring (Bairlein, 1991a). Since these reserves are required during rather precisely defined times of the year, and since carrying large fat reserves has obvious costs and risks (e.g., Witter and Cuthill, 1993; Kullberg *et al.*, 1996; Fransson and Weber, 1997; Kullberg, 1998; Lind *et al.*, 1999), appropriate timing and the amount of deposition are of utmost significance. Therefore, long-distance migrants are equipped with a sophisticated timing system comprising both endogenous and exogenous components (Bairlein and Gwinner, 1994; Berthold, 1996). Consequently, these species may be the most affected ones by changing environmental conditions at their stopover sites, as they often rely on a few particular stopover sites while medium- and short-distance migrants may be more flexible in their decision where to rest.

In many of these long-distance migrating species, migratory fuelling happens immediately before crossing ecological barriers (e.g., Moore and Kerlinger, 1987; Biebach, 1990, 1995; Bairlein, 1991a, 1998; Yong *et al.*, 1998; Schaub and Jenni, 2000a,b, 2001a,b; Ottosson *et al.*, 2002). Shortly after an obstacle, migrants need sites to recover from the long flights and to prepare for the continuation of migration (e.g., Biebach and Bauchinger, 2003). At all these sites, fuelling migrants largely rely on appropriate habitats (Bairlein, 1981, 1992) and on good feeding opportunities both in terms of quantity and quality (Bairlein, 1991b, 1998, 2002, 2003; Dierschke and Delingat, 2001; Jenni-Eiermann and Jenni, 2003). Predictability of these conditions is a basic feature in the birds' "strategies" to migrate (Alerstam and Lindström, 1990; Weber, 1999; Weber *et al.*, 1999). But predictability of these conditions and of the spatial distribution of habitats for stopover may change owing to climate change. This may also be different along different flyways due to pronounced geographical variation in the effects of climate change and population-specific different fuelling patterns along different flyways, as shown, for example, in garden warblers (Figure 1; Bairlein, 1991a) or barn swallows (Rubolini *et al.*, 2002).

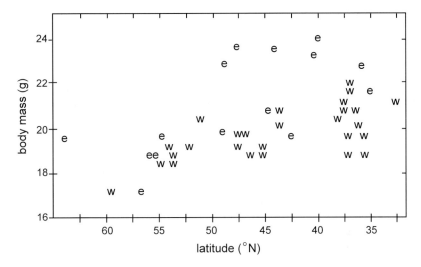

Figure 1   Geographical variation in the pattern of autumn migratory fuelling in garden warblers. Birds migration along the eastern European flyway (e) put on fuel at more northerly sites than birds along the western flyway (w). Redrawn after Bairlein (1991a).

V. PLASTICITY IN MIGRATORY PERFORMANCE: DOES IT MATTER?

In certain migrant species, the temporal and spatial control of migration is largely by innate mechanisms that determine timing of migration, distance to migrate, direction of migration, orientation and fuelling (for review, see Berthold, 1996; Bairlein *et al.*, 2002). However, there may be enough genetic variation that birds can accommodate by selection (Pulido and Berthold, 2003; A.P. Møller, manuscript), as well as individual birds show some phenotypic plasticity to adjust their innate programmes for adverse influences on migratory performance (Schindler *et al.*, 1981; Biebach, 1985; Gwinner *et al.*, 1985, 1988, 1992; Sutherland, 1998; Lindström and Agrell, 1999). Experimentally induced adverse weather conditions or artificial food deprivation in caged warblers led to an increase in duration and intensity of nocturnal migratory activity and even to a re-induction of nocturnal migratory activity after the termination of spontaneous autumn migratory restlessness showing that the innate migratory programs are susceptible to modifying external cues (Figure 2).

Migrants seem to be able to counteract deteriorating habitat and food conditions. Consequently, an extension of the Sahara belt due to climate warming may not necessarily cause an adverse impact on the migratory performance of birds. Rather, these migrants are able to enhance migratory activity while facing hostile areas and to reactivate migratory activity when the

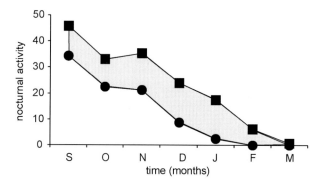

Figure 2 Food-deprived captive garden warblers (squares) increase nocturnal migratory activity as compared to birds fed *ad libitum* (dots). Redrawn after Gwinner *et al.* (1988).

food supply is deteriorating (Gwinner *et al.*, 1988). However, the distances they can travel by these compensatory mechanisms are difficult to estimate, and there may be energetic limits.

VI. ENERGY STORES AND MIGRATION PERFORMANCE

In free-living migrants, the flight distances or durations are limited by the amount of fuel aboard (e.g., Pennycuick, 1989; Biebach, 1992; Wikelski *et al.*, 2003). Consequently, the feeding conditions at stopover sites where the major fuel stores are accumulated are of utmost importance for the temporal design of the entire travel and the overall migration speed. Habitats for feeding may become scarcer and they may deteriorate due to drought or vegetation shifts. The abundance of food may become lower as well as its composition may change in space and time.

Lower food supply affects the carrying capacity of stopover sites (Sutherland and Goss-Custard, 1991; Alerstam *et al.*, 1992; Ottich and Dierschke, 2003), as stopover sites typically attract many migrants in short periods. Pronounced adverse effects can be expected by a mismatch between the temporal course of fuelling and the availability of food. While the return to breeding grounds or the departure to wintering grounds in migrating birds may be initiated by endogenous mechanisms or photoperiod, the availability of food depends mainly on temperature and precipitation. Consequently, advancing vegetation and earlier appearance of insects in spring due to warmer and wetter spring conditions may lead to a mismatch with the passage of migrants. That mismatch may partly explain observations by Møller (manuscript) that Danish barn swallows do not only arrive earlier in years with adverse spring conditions in

Northern Africa (low NDVI) but also fewer individuals in better body condition arrive and subsequent mortality rates increase.

For staging migrants, low food supply affects the birds' ability to accumulate appropriate energy stores for subsequent migration. Insufficient food supply for refuelling is shown to trigger stopover behaviour and departure time (e.g., Alerstam and Hedenström, 1998; Ottich and Dierschke, 2003). It also has a strong influence on orientation and selection of direction (Bäckman et al., 1997; Sandberg, 2003). In all species tested, fat birds oriented towards the seasonally appropriate directions while lean birds showed reverse migration.

However, migrants seeking a site for landing seem to adjust their decision according to their energy stores and the expected food supply. At Saharan oases, lean birds avoid to land at small places with less food (Figure 3; Biebach et al., 1986; Bairlein, 1992). On the island of Helgoland, German North Sea, body mass

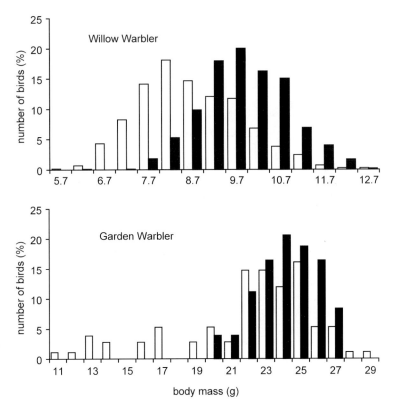

Figure 3 Trans-Sahara migrating birds landing at large oases (open bars) are lighter than birds at small oases (filled bars). Top: Willow warblers in the Egyptian desert (redrawn after Biebach et al., 1986). Bottom: Garden warbler in the Algerian Sahara. Redrawn after Bairlein (1987).

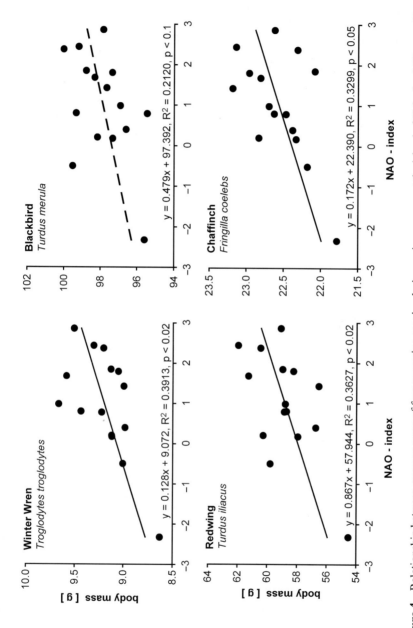

Figure 4 Relationship between average mass of four passerine species during spring passage at the island of Helgoland (German North Sea) and the North Atlantic Oscillation Index (Hüppop and Hüppop, 2003).

40

of 16 passerine migrant species with sufficient sample size trapped during spring passage show a close relationship with the NAO-index (Hüppop and Hüppop, 2003; Figure 4). Birds may avoid deteriorating habitats in Northern Africa and Southern Europe or may be forced to leave earlier, as do barn swallows, but face better feeding conditions northbound in the warmer and wetter environmental conditions in Western Europe during positive NAO-index years, which are likely to enhance abundance of food for staging migrants. Food shortage at deteriorating sites of stopover may even increase the fuel deposition rate. In captive garden warblers, short-term food restriction accelerated daily body mass gain (Figure 5; Totzke *et al.*, 2000). Birds also may face more favourable south-westerly tail wind conditions thus lowering flight costs. Moreover, warmer temperatures in spring may reduce thermoregulatory costs at the stopover sites (Scheiffarth, 2003; Wikelski *et al.*, 2003), which is likely to influence both fuel deposition and timing of departure.

Another kind of climate change induced mismatch between migratory timing and conditions for fuelling of migrants at stopover sites may be faced by arctic geese and swans. As herbivores, their timing of spring migration largely depends on the availability of high-quality forage plants at a few staging sites (e.g., Prop and Black, 1998; Stahl, 2001). Geese, for example, move north following a succession of sites at a time when the local food quality peaks

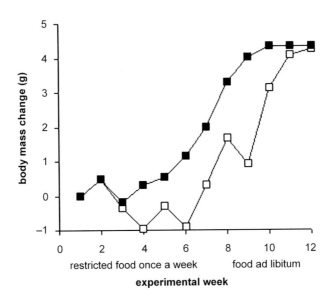

Figure 5   Short-term food-deprived captive garden warblers (filled symbols; week 1–7) put on migratory body mass faster than continuously *ad libitum* fed control birds (open symbols). Redrawn after Totzke *et al.* (2000).

("green wave"; Owen, 1980). Consequently, changes in plant phenology due to climate change will affect the timing of geese migration. Regional variation in the amount and speed of climate change along the migratory routes is thus very likely to lead to asynchrony between the traditional migratory schedule and the availability of forage for fuelling. However, appropriate fuelling during staging is crucial for subsequent reproductive success (e.g., Ebbinge and Spaans, 1995; Prop and Black, 1998).

VII. OUTLOOK

If the magnitude of climate change further increases, the number of systems affected and the geographical extent will increase (IPCC, 2001; Berry et al., 2003). But habitats may show specific responses (e.g., Berry et al., 2003), and species may differ in their response so that some may win while others lose (e.g., Harrison et al., 2003b). However, predicting and assessing the impact of these changes on migratory birds is hindered by the lack of detailed studies and the uncertainty in climate models, especially with respect to regional changes. The few data presented here reveal pronounced species-specific differences in the influence of global warming-mediated ecological factors on migratory performance in birds depending on the birds' migratory habits. Owing to the geographical variation of climate change, namely of NAO effects (Ottersen et al., 2001; Visbeck et al., 2001), western migrants may face more severe changes than eastern migrants, as the western European–African flyway is likely to be more affected than eastern regions with deteriorating effects in the southwest and improving ecological condition further to the north. Long-distance migrants may be more vulnerable to these changes because they may rely more on innate mechanisms in the control of the temporal and spatial course of their migrations than short and medium distance migrants where exogenous factors may play a larger role. Moreover, birds migrating in many short hops may be less susceptible to changes in the distribution of habitats for fuelling than birds migrating in few long flights with stopovers and fuelling at a few sites (e.g., Piersma, 1987). Terrestrial birds (Berry et al., 2003) and birds using freshwater habitats (Dawson et al., 2003) may be more affected than coastal migrants, although the latter may face substantial effects of sea level rise (Neuhaus et al., 2001). Sea level rise changes inter-tidal habitats and morphology of estuaries and affects availability of habitats and food for waders because landward extension of salt marshes is restricted by dikes and other sea defences in almost every area. This not only changes habitat for wintering and stopover shorebird species (Sutherland and Goss-Custard, 1991; Lindström and Agrell, 1999; Galbraith et al., 2002; Austin and Refisch, 2003), but may also affect timing of migration and subsequent reproduction as the latter mainly depends on a precise timing to make full use of the summer peak in food abundance, which is of utmost importance in arctic breeding species (e.g., Ankney and McInnes, 1978; Schmutz and Ely, 1999; Piersma, 2003; Schekkerman et al., 2003).

Further research is needed on the relationship between climate change and associated changes in habitats and food resources for migrants. Modelling natural resource responses to climate change (Harrison et al., 2003a) to evaluate the impacts on the natural systems and on the distribution of habitats for fuelling of migrants is one way to understand and predict the current and future effects of climate change on wildlife (e.g., Berry et al., 2003; Dawson et al., 2003; Harrison et al., 2003b). Considering migrants, it is particularly important to explore the relationship between climate change and associated ecological changes across broad geographic regions and various flyways. The other approach is to gather more detailed data about stopover and fuelling conditions of migrants in areas, which are predicted as the ones with most severe changes due to climate change, e.g., southern Spain, Northern Africa (Bairlein, 1988, 1997; Rguibi et al., 2003), the Sahel (Ottosson et al., 2002), and presumably the Wadden Sea (Piersma, 2003). Moreover, the conditions at the winter grounds should be explored in more detail. In Southern Africa, vegetation productivity increases during high NAO-index years, which is likely to increase abundance of insects for wintering migrants (Forchhammer et al., 2002). Improved feeding conditions during winter may cause earlier departures because of an increased rate of body condition gain (Marra et al., 1998). Improved body condition may not only affect migratory performance. In 14 species of birds, Møller and Erritzøe (2003) found close positive relationships between body condition and spleen mass, and between NAO-index and spleen mass. Thus, climate change as indicated by, for example, the NAO-index is also affecting immune defence organs and thus parasite-mediated natural selection. On the other hand, it has been shown that climate change may decrease primary productivity and thus fish stocks in aquatic ecosystems of Africa (O'Reilly et al., 2003) that may have considerable consequences for migratory fish-eating birds.

In temperate regions, warmer conditions may induce birds to winter further north than usual (for examples of different winter distribution between cold and mild winters see, e.g., Meltofte et al., 1994). Due to lower thermoregulatory costs, energetic cost for life is in most cases lower in warmer climates (Bairlein, 1993). Arctic breeding red knots Calidris canutus, for example, spent 1.50 W for maintenance metabolism when wintering in tropical western Africa but would have to spend 2.93 W in January in the Netherlands (Wiersma and Piersma, 1994). In bar-tailed godwit Limosa lapponica wintering along the North Sea coasts, winter distribution is shaped by thermoregulatory requirements (Scheiffarth, 2003). Consequently, climate change with increasing temperatures in Europe will reduce thermoregulatory costs, probably inducing more arctic shorebirds to stay over winter in the Wadden Sea with substantial influences on their migratory and pre-breeding fattening scheme (Piersma, 2003; Scheiffarth, 2003).

However, in addition to climate, other factors of environmental change, in particular, loss of habitats and food resources due to human activities, must always be assessed simultaneously in order to separate climate effects from the many other impacts to evaluate proper climate impact models and predictions.

REFERENCES

Alerstam, T. and Hedenström, A. (1998) *J. Avian Biol.* **29**, 337–636.

Alerstam, T. and Lindström, Å. (1990) In: *Bird Migration. Physiology and Ecophysiology* (Ed. by E. Gwinner), pp. 331–351. Springer-Verlag, Berlin.

Alerstam, T., Gudmundsson, G.A. and Johannesson, K. (1992) *Oikos* **65**, 179–189.

Ankney, C.D. and MacInnes, C.D. (1978) *Auk* **95**, 459–471.

Austin, G.E. and Rehfisch, M.M. (2003) *J. Nat. Conserv.* **11**, 43–58.

Bäckman, J., Pettersson, J. and Sandberg, R. (1997) *Ethology* **103**, 247–256.

Bairlein, F. (1981) *Ökol. Vögel* **3**, 7–137.

Bairlein, F. (1987) *Ringing & Migration* **8**, 59–72.

Bairlein, F. (1988) *Vogelwarte* **34**, 237–248.

Bairlein, F. (1991a) *Vogelwarte* **36**, 48–61.

Bairlein, F. (1991b) *Proc. XX. Int. Orn. Congr.*, 2149–2158.

Bairlein, F. (1992) In: *Ecology and Conservation of Neotropical Migrant Landbirds* (Ed. by J.M. Hagan and D.W. Johnston), pp. 356–369. Smithsonian, Washington.

Bairlein, F. (1993) *Proc. VIII Pan-Afr. Orn. Congr.*, 571–578.

Bairlein, F. (1997) *Jber. Institut Vogelforschung* **3**, 17–18.

Bairlein, F. (1998) *Biol. Cons. Fauna* **102**, 13–27.

Bairlein, F. (2002) *Naturwissenschaften* **89**, 1–10.

Bairlein, F. (2003) In: *Avian Migrations* (Ed. by P. Berthold, E. Gwinner and E. Sonnenschein), pp. 293–306. Springer-Verlag, Berlin.

Bairlein, F., Elkins, N. and Evans, P. (2002) In: *The Migration Atlas. Movements of the Birds of Britain and Ireland* (Ed. by C. Wernham, M. Toms, J. Marchant, J. Clark, G. Shiriwardena and S. Baillie), pp. 23–43. T. and A.D. Poyser, London.

Bairlein, F. and Gwinner, E. (1994) *Annu. Rev. Nutr.* **14**, 187–215.

Bairlein, F. and Winkel, W. (2001) In: *Climate of the 21st Century: Changes and Risks* (Ed. by J.L. Lozán, H. Graßl and P. Hupfer), pp. 278–282. Wissenschaftliche Auswertungen, Hamburg.

Berry, P.M., Dawson, T.P., Harrison, P.A., Pearson, R. and Butt, N. (2003) *J. Nat. Conserv.* **11**, 15–23.

Berthold, P. (1996) *Control of Bird Migration*. Chapman & Hall, London.

Berthold, P., Gwinner, E., and Sonnenschein, E. (Eds.) (2003) *Avian Migration*. Springer-Verlag, Berlin.

Bibby, C.J. and Green, R.E. (1981) *Ornis Scand.* **12**, 1–12.

Biebach, H. (1985) *Experientia* **41**, 695–697.

Biebach, H. (1990) In: *Bird Migration: Physiology and Ecophysiology* (Ed. by E. Gwinner), pp. 399–412. Springer-Verlag, Berlin.

Biebach, H. (1992) *Ibis* **134**, 47–54.

Biebach, H. (1995) *Israel J. Zool.* **41**, 387–392.

Biebach, H. (1996) In: *Avian Energetics and Nutritional Ecology* (Ed. by C. Carey), pp. 280–323. Chapman & Hall, New York.

Biebach, H. and Bauchinger, U. (2003) In: *Avian Migrations* (Ed. by P. Berthold, E. Gwinner and E. Sonnenschein), pp. 269–280. Springer-Verlag, Berlin.

Biebach, H., Friedrich, W. and Heine, G. (1986) *Oecologia* **69**, 370–379.

Buttler, C.J. (2003) *Ibis* **145**, 484–495.

Cubasch, U. (2001) In: *Climate of the 21st century: Changes and Risks* (Ed. by J.L. Lozán, H. Graßl and P. Hupfer), pp. 173–179. Wissenschaftliche Auswertungen, Hamburg.

Dawson, T.P., Berry, P.M. and Kampa, E. (2003) *J. Nat. Conserv.* **11**, 25–30.

Dierschke, V. and Delingat, J. (2001) *Behav. Ecol. Sociobiol.* **50**, 535–545.

Ebbinge, B.S. and Spaans, B. (1995) *J. Avian Biol.* **26**, 105–113.

Forchhammer, M.C., Post, E. and Stenseth, N.C. (2002) *J. Anim. Ecol.* **71**, 1002–1014.

Fransson, T. and Pettersson, J. (2001) *Swedish Bird Ringing Atlas*, Vol. 1, Naturhistoriska Riksmuseet Sveriges and Onritologiska Förening.

Fransson, T. and Weber, T.P. (1997) *Behav. Ecol. Sociobiol.* **41**, 75–80.

Galbraith, H., Jones, R., Park, R., Clogh, J., Herrod-Julius, S., Harrington, B. and Page, G. (2002) *Waterbirds* **25**, 173–183.

Gwinner, E. (1990) *Bird Migration. Physiology and Ecophysiology*. Springer-Verlag, Heidelberg.

Gwinner, E., Biebach, H. and Kries, I.V. (1985) *Naturwissenschaften* **72**, 51–52.

Gwinner, E., Schwabl, H. and Schwabl-Benzinger, I. (1988) *Oecologia* **77**, 321–326.

Gwinner, E., Schwabl, H. and Schwabl-Benzinger, I. (1992) *Ornis Scand.* **23**, 264–270.

Harrison, P.A., Berry, P.M. and Dawson, T.P. (2003a) *J. Nat. Conserv.* **11**, 3–4.

Harrison, P.A., Vanhinsbergh, D.P., Fuller, R.J. and Berry, P.M. (2003b) *J. Nat. Conserv.* **11**, 31–42.

Hüppop, O. and Hüppop, K. (2003) *Proc. R. Soc. Lond. B* **270**, 233–240.

Hurrell, J.W. (1995) *Science* **269**, 676–679.

IPCC (2001) Climate Change 2001. Cambridge University Press, Cambridge.

Jenni, L. and Kéry, M. (2003) *Proc. R. Soc. Lond. B* **270**, 1467–1471.

Jenni-Eiermann, S. and Jenni, L. (2003) In: *Avian Migrations* (Ed. by P. Berthold, E. Gwinner and E. Sonnenschein), pp. 293–306. Springer, Berlin.

Kullberg, C. (1998) *Anim. Behav.* **56**, 227–233.

Kullberg, C., Fransson, T. and Jakobson, S. (1996) *Proc. R. Soc. Lond. B* **263**, 1671–1675.

Lind, J., Fransson, T., Jakobsson, S. and Kullberg, C. (1999) *Behav. Ecol. Sociobiol.* **46**, 65–70.

Lindbladh, M., Bradshaw, R. and Holmqvist, B.H. (2000) *J. Ecol.* **88**, 113–128.

Lindström, Å. (2003) In: *Avian Migrations* (Ed. by P. Berthold, E. Gwinner and E. Sonnenschein), pp. 307–320. Springer, Berlin.

Lindström, Å. and Agrell, J. (1999) *Ecol. Bull.* **47**, 145–159.

Marra, P.P., Hobson, K. and Holmes, R.T. (1998) *Science* **282**, 1884–1886.

Meltofte, H., Blew, J., Frikke, J., Rösner, H.-U. and Smit, C.J. (1994) *Numbers and Distribution of Waterbirds in the Wadden Sea*. CWSS, Wilhelmshaven.

Møller, A.P. and Erritzøe, J. (2003) *Oecologia* **136**, 621–626.

Moore, F. and Kerlinger, P. (1987) *Oecologia* **74**, 47–54.

Neuhaus, R., Dijkema, K.S. and Reinke, H.-D. (2001) In: *Climate of the 21st Century: Changes and Risks* (Ed. by J.L. Lozán, H. Graßl and P. Hupfer), pp. 311–314. Wissenschaftliche Auswertungen, Hamburg.

Oba, G., Post, E. and Stenseth, N.C. (2001) *Global Change Biol.* **7**, 241–246.

O'Reilly, C.M., Alin, S.R., Plisnier, P.-D., Cohen, A.S. and Mckee, B.A. (2003) *Nature* **424**, 766–768.

Ottersen, G., Planque, B., Belgrano, A., Post, E., Reid, P.C. and Stenseth, N.C. (2001) *Oecologia* **128**, 1–14.

Ottich, I. and Dierschke, V. (2003) *J. Ornithol.* **144**, 307–316.

Ottosson, U., Bairlein, F. and Hjort, C. (2002) *Vogelwarte* **41**, 249–262.

Owen, M. (1980) *Wild Geese of the World*. Batsford Ltd, London.

Parmesan, C. and Yohe, G. (2003) *Nature* **421**, 37–42.

Pennycuick, C.J. (Ed.) (1989) *Bird Flight Performance. A Practical Calculation Manual*. Oxford University Press, Oxford.

Peñuelas, J. and Boada, M. (2003) *Global Change Biology* **9**, 131.

Piersma, T. (1987) *Limosa* **60**, 185–194.

Piersma, T. (2003) In: *Warnignale Nordsee und Wattenmeer* (Ed. by J.L. Lozán, E. Rachor, K. Reise, J. Sündermann and H. von Westernhagen), pp. 176–181. Wissenschaftliche Auswertungen, Hamburg.

Pilardeaux, B. and Schulz-Baldes, M. (2001) In: *Climate of the 21st Century: Changes and Risks* (Ed. by J.L. Lozán, H. Graßl and P. Hupfer), pp. 232–236. Wissenschaftliche Auswertungen, Hamburg.

Prop, J. and Black, J.M. (1998) Norsk Polarinst. *Skrifter* **200**, 175–193.

Pulido, F. and Berthold, P. (2003) In: *Avian Migrations* (Ed. by P. Berthold, E. Gwinner and E. Sonnenschein), pp. 53–77. Springer, Berlin.

Rguibi Idrissi, H., Julliard, R. and Bairlein, F. (2003) *Ibis* **145**, 650–656.

Root, T.L., Price, J.T., Hall, K.R., Schneider, S.H., Rosenzweig, C. and Pounds, J.A. (2003) *Nature* **421**, 57–60.

Rubolini, D., Gardiazabal Pastor, A., Pilastro, A. and Spina, F. (2002) *J. Avian Biol.* **33**, 15–22.

Sandberg, R. (2003) In: *Avian Migrations* (Ed. by P. Berthold, E. Gwinner and E. Sonnenschein), pp. 515–525. Springer, Berlin.

Schaub, M. and Jenni, L. (2000a) *Oecologia* **122**, 306–317.

Schaub, M. and Jenni, L. (2000b) *J. Ornithol.* **141**, 441–460.

Schaub, M. and Jenni, L. (2001a) *Funct. Ecol.* **15**, 584–594.

Schaub, M. and Jenni, L. (2001b) *Oecologia* **128**, 217–227.

Scheiffarth, G. (2003) Born to fly—migratory strategies and stopover ecology in the European Wadden Sea of a long-distance migrants, the Bar-tailed Godwit (*Limosa lapponica*), Doctoral Thesis, University of Oldenburg.

Schekkerman, H., Tulp, I., Piersma, T. and Visser, G.H. (2003) *Oecologia* **134**, 332–342.

Schindler, J., Berthold, P. and Bairlein, F. (1981) *Vogelwarte* **31**, 33–44.

Schmutz, J.A. and Ely, C.R. (1999) *J. Wildl. Manage.* **63**, 1239–1249.

Stahl, J. (2001) Limits to the co-occurrence of avian herbivores, Doctoral Thesis, University of Groningen.

Strode, P.K. (2003) *Global Change Biol.* **9**, 1137–1144.

Sutherland, W. (1998) *J. Avian Biol.* **29**, 441–448.

Sutherland, W.J. and Goss-Custard, J.D. (1991) *Acta XX Congr. Int. Ornithol.*, 2199–2207.

Totzke, U., Hübinger, A., Dittami, J. and Bairlein, F. (2000) *J. Comp. Physiol. B* **170**, 627–631.

Visbeck, M.H., Hurrell, J.W., Polvani, L. and Cullen, H.M. (2001) *Proc. Natl Acad. Sci. USA* **98**, 12876–12877.

Walther, G.-R., Post, E., Convey, A., Menzel, A., Parmesan, C., Beebee, T.J.C., Fromentin, J.-M., Hoegh-Guldberg, O. and Bairlein, F. (2002) *Nature* **416**, 389–395.

Wang, G. (2003) *Global Change Biology* **9**, 493–499.

Weber, T.P. (1999) *J. Theor. Biol.* **199**, 415–424.

Weber, T.P., Fransson, T. and Houston, A.I. (1999) *Behav. Ecol. Sociobiol.* **46**, 280–286.

Wernham, C., Toms, M., Marchant, J., Clark, J., Siriwardena, G. and Baillie, S. (2002) *The Migration Atlas. Movements of the Birds of Britain and Ireland.* British Trust for Ornithology.

Wiersma, P. and Piersma, T. (1994) *Condor* **96**, 257–279.

Wikelski, M., Tarlow, E.M., Raim, A., Diehl, R.H., Larkin, R.P. and Visser, G.H. (2003) *Nature* **423**, 704.

Witter, M.S. and Cuthill, I.C. (1993) *Phil. Trans. R. Soc. Lond. B* **340**, 73–92.

Yong, W., Finch, D.M., Moore, F.R. and Kelly, J.F. (1998) *Auk* **115**, 829–842.

Zahn, R. (2003) *Nature* **421**, 324–325.

Zink, G. (1973, 1975, 1981, 1985) *Der Zug Europäischer Singvögel: ein Atlas der Wiederfunde Beringter Vögel,* Lfg. 1-4. AULA-Verlag, Wiesbaden.

Zink, G. and Bairlein, F. (1995) *Der Zug Europäischer Singvögel: ein Atlas der Wiederfunde Beringter Vögel,* Vol. 3. AULA-Verlag, Wiesbaden.

Using Large-Scale Data from Ringed Birds for the Investigation of Effects of Climate Change on Migrating Birds: Pitfalls and Prospects

WOLFGANG FIEDLER,* FRANZ BAIRLEIN AND ULRICH KÖPPEN

I. SUMMARY

Ringing recovery datasets at ringing centres cover large areas and long time-spans, and, therefore, they are interesting for investigating long-term changes in bird migration. However, heterogeneity of data in terms of ringing activity, recapture and re-sighting efforts and recovery probabilities as well as reporting probabilities of recoveries on a temporal and spatial scale are major obstacles to such analyses.

Among 30 species of short-distance or partial migrants from Germany we found evidence for significantly increased proportions of winter recoveries within a distance less than 100 km in recent years in nine species. We found evidence for reduced mean recovery distances in eight species and evidence for increased mean recovery distances in five species. Increases in recovery distances might result from changes in reporting probabilities of hunted birds,

---

E-mail address: fiedler@orn.mpg.de

ADVANCES IN ECOLOGICAL RESEARCH, VOL. 35
0065-2504/04 $35.00 DOI 10.1016/S0065-2504(04)35003-8

density-dependent changes and probably other reasons. A tendency towards wintering at higher latitudes in recent years was found in 10 species while mean latitudes of winter recoveries moved southwards in three species. These results are mainly consistent with studies from other European regions.

Changes in geographical distribution of recoveries may be interpreted from a global warming perspective, although only one study (Soutullo, 2003) has so far demonstrated direct relationships (although inconsistent) between migration distance (wintering latitude) and climate variables. Since other factors such as changes in land use, winter feeding, availability of garbage dumps and other environmental changes may affect the position of wintering areas, these changes have to be taken into account when changes in migratory behaviour are analysed.

II. INTRODUCTION

In several countries scientific bird ringing has been in constant use for over 100 years. In addition to the databases that were kept by the conductors of the many different studies that apply individual marking of birds, basic ringing and recovery data are also stored for a wider region and over a longer time period at ringing centres. Earlier studies to detect changes in migratory behaviour of waterfowl on a national scale have been performed by Møller (1981), Dagys (2001), Remisiewicz (2001) and Švažas et al. (2001). Recently, Siriwardena and Wernham (2002) used the recovery database of birds from Britain and Ireland to calculate indices of migratory tendency which can be used for a variety of analyses, including detection of changes in migratory behaviour. Soutullo (2003) was the first to show relationships between mean wintering latitude and climate variables in this dataset.

We analysed selected data from three German ringing centres "Helgoland" (Northern Germany, active since 1909), "Hiddensee" (area of the former German Democratic Republic, active since 1964) and "Radolfzell" (Southern Germany, active in the current region since 1947) to explore the value of German recovery data for investigating temporal changes in bird migration behaviour. Although heterogeneity of the data in terms of ringing activity, recapture, re-sighting effort, recovery probabilities or reporting probabilities of recoveries on a temporal and spatial scale are major problems for these types of long-term analyses, the data offer promising possibilities: (1) ringing and recovery databases at ringing centres cover larger areas and longer time-spans than most single studies, (2) in contrast to pure observations and bird counts, individuals with deviant behaviour (like wintering at northern latitudes by migrants) can be assigned to distinct populations, and (3) the datasets are rapidly available in standardised, electronic format (Speek et al., 2001). Thanks to the coordinating efforts of the European Union for Bird Ringing (EURING) analyses of changes in migration behaviour of some species might cover many decades and large parts of Europe (e.g., white stork *Ciconia ciconia*, Fiedler, 2001). After presenting some general ideas on the use of ringing data for analyses of effects of climate change on bird migration, we use the German datasets to investigate prospects and pitfalls in using data from bird ringing databases for analysing the effects of climate change on migrating birds. Finally, we discuss a few, exciting analyses that have already been done on this subject.

III. CHANGES IN MIGRATION BEHAVIOUR

A. Types of Changes

The currently reported changes in bird migration behaviour relative to climate change can be ascribed to four main processes (Fiedler, 2003): (1) changes in migration status (like the proportion of non-migrants in a population), (2) changes in migration distance, (3) changes in migration direction, and (4) changes in timing and speed of migration.

Migration can be regarded as a function of the position of the individual in space and time (Figure 1). Changes in migration by individuals will appear as deviations from the normal pattern (thin, dotted and shaded lines), which represents the vast majority of a population (thick solid line), by appearing at

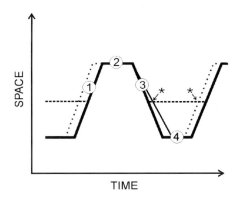

Figure 1 Schematic representation of migration with a spatial and temporal axis. 1: migration to winter quarters; 2: stay in winter quarter; 3: migration to breeding grounds; and 4: stay at the breeding grounds. Thick solid line: starting pattern; dotted line: example of earlier migration with unchanged migration speed to the breeding grounds; shaded line: example of a change in winter location with unchanged migration speed; thin solid line: example of a change in migration speed (slower migration speed to winter quarters). Stars, see text.

unusual places or during unusual times of the year. By analysing the spatial and temporal information in ringing recovery data we can: (1) assess changes over time in presence at a distinct area (like occurrence of birds during times when they were previously absent, earlier arrival or longer stays), (2) we can investigate changes in recovery distances calculated from coordinate information of ringing and recovery locations or at mean recovery latitudes at certain times of the year, and (3) we can make use of standardised trapping activities (like ringing field stations), which can quantify the passage of birds through a distinct area during a specific period.

Given the potential changes shown in Figure 1, it is obvious that it will often be impossible to distinguish between different types of changes. The stars mark the situations where we cannot decide whether we face a change in location of the winter quarters or in the timing or the speed of migration. There might be only rare sufficient additional information from other marked individuals of the same population to draw a reliable conclusion about the observed type of change.

One obvious way to assess reactions of migrating birds to climatic change is: (1) to test for trends or more complicated time-dependent patterns in migration behaviour, (2) then eliminate detected changes that have occurred for reasons other than climate change, and (3) finally investigate co-variation with climate data. In addition, correlations between different climate variables and recovery distances in single years can be investigated. The first test risks neglecting relationships between climate and migration due to the lack of any obvious trend in migration behaviour, while the second step might be of major importance due to the various pitfalls arising from the use of large-scale ringing recovery data.

B. Changes in the Presence of Individuals at Distinct Times

A combination of field observations of the increasing numbers of wintering birds of a particular, formerly fully migratory species and the increasing number of local winter recoveries are particularly strong indication of changes in migratory pattern. Since there is no evidence that winter recovery and reporting probabilities have changed disproportionately to other recoveries in Europe, we can expect unbiased results if we control: (1) the changes in causes of death, (2) the changes in ringing activity, and (3) the changes in recapture and re-sighting activity.

Hunting is in many species still the major cause of recovery among ringed birds that were found dead, and regional or national changes in hunting regulations and in reporting behaviour by hunters must be considered. Other frequently reported causes of death like electrocution in white storks (Fiedler and Wissner, 1989) may also dramatically influence patterns of temporal and geographic distribution of recoveries in single species. In our analysis, we separated birds reported as "shot" or "hunted" from those with other recovery circumstances and used all individuals reported as recovered for "unknown" reasons as a third group to assess the extent of "dishonest" reports of hunted birds.

Temporal changes in ringing activities may bias patterns of recoveries since increasing ringing activity in southern regions and decreasing activity in the north may cause an apparent increase in winter recoveries close to the breeding grounds for partial migrants. Therefore, we made frequency-scatter plots of latitude of ringing in relation to year for all species investigated. To avoid influences of systematic changes in recapture and re-sighting activity, we excluded all re-sightings (recaptures of a live bird) and all recoveries with identical ringing and recovery coordinates, since this often indicates activities in study plots where increased efforts to obtain local recaptures are likely. For the same reason recoveries coded as being "trapped and released by a ringer" were excluded because recapture activities of ringers may also imply temporally and locally highly increased recovery probabilities. Wernham and Siriwardena (2002) also discuss these potential pitfalls in detail.

From the German bird ringing recovery dataset, we selected 40 species of short or medium-distance migrants with sufficient sample sizes. To select only birds of known population origin, we restricted the analysis to individuals ringed as nestlings or as adults during the breeding season as defined for each species according to the literature (Glutz von Blotzheim, 1966–1997). The winter period was defined for all species as November–February. Inaccurate spatial and temporal information (coordinate accuracy $<1°$, date accuracy <2 weeks) also led to exclusion of recoveries. Among the remaining 30 species with sufficient sample sizes (more than 30 winter recoveries) we found evidence for an increased proportion of winter recoveries at a distance less than 100 km in 9 species (Table 1). This is much more than the 1.5 species that are expected to show the change in the predicted direction with a significance level of 5%.

Figure 2 shows an example data for migration distance during the last 80 years for the black redstart *Phoenicurus ochruros*. Recoveries occur at a fairly constant number with the exception of small number during World War II and since the 1990s, when ringing effort was much reduced (Figure 2a). The first regular local winter recoveries occurred in 1954, while most birds were still recovered at distances of 800–2000 km in south-westerly directions during winter. The local wintering birds were not recruited among the northern and north-easternmost breeding birds (Figure 2b).

Despite the availability of numerous observations and bird reports, so far only few studies used bird ringing data to identify the origin of wintering individuals (e.g., Adriaensen *et al.* (1993) for Dutch great crested grebes *Podiceps cristatus*, Petersen (1984) for Danish herring gulls *Larus argentatus*, George (1995) for German red kites *Milvus milvus*, Loonen and De Vries (1995) for Dutch graylag geese *Anser anser*). However,

Table 1 Observed changes in migration behaviour in 30 breeding bird species from Germany on the basis of ringing recovery data

Species	Sample size[a]	Winter recoveries[b]	Distance (km)[c]			Latitude[d]			Period[e]
			β	SE	p	β	SE	p	
Phalacrocorax carbo	100	X	−0.09	647.8	0.379	0.078	5.96	0.445	1965–1996
Ardea cinerea	358		0.057	685.9	0.281	−0.01	5.3	0.847	1925–2000
Tadorna tadorna	28	X	−0.51	250.6	0.005	0.429	1.69	0.023	1950–1973
Anas plathyrhynchos	201		0.088	369.0	0.214	0.072	2.56	0.309	1929–2000
Milvus milvus	207	X	−0.07	688.4	0.294	0.81	4.76	0.238	1952–2001
Buteo buteo	387	X	−0.3	174.2	0.616	0.152	1.59	0.003	1960–1986
Accipiter nisus	490	X	−0.27	339.5	<0.001	0.215	2.72	<0.001	1922–2002
Falco tinnunculus	845		−0.06	353.5	0.075	0.016	3.19	0.638	1928–2001
Fulica atra	190	X	−0.13	350.0	0.071	−0.05	3.68	0.527	1954–1991
Haematopus ostralegus	223		0.024	342.7	0.725	0	2.4	0.997	1927–2001
Charadrius hiaticula	34		−0.31	708.9	0.074	0.313	5.54	0.072	1953–1986
Vanellus vanellus	557		−0.27	524.4	<0.001	0.2	4.16	<0.001	1950–2001
Larus ridibundus	531		−0.08	570.4	0.064	0.111	4.82	0.01	1928–2002
Larus canus	364		−0.11	449.9	0.038	0.083	3.31	0.114	1928–1994
Columba oenas	170		0.202	371.5	0.008	−0.12	2.9	0.125	1978–2001

Species	n[a]	[b]	β[c]	SE[c]	p	β[d]	SE[d]	p	Period[e]
Columba palumbus	152		0.348	458.4	<0.001	−0.32	3.66	<0.001	1922–1999
Motacilla alba	58		0.215	500.3	0.105	−0.27	3.59	0.044	1950–1992
Prunella modularis	28		0.06	491.7	0.76	−0.07	3.62	0.711	1934–1979
Phoenicurus ochruros	72	X	−0.33	652.1	0.004	0.325	4.82	0.005	1960–1990
Turdus merula	226		−0.14	417.1	0.038	0.279	4.65	<0.001	1954–2002
Turdus philomelos	255		0.234	511.1	<0.001	−0.04	4.07	0.48	1924–2000
Sturnus vulgaris	156		0.572	485.1	<0.001	−0.5	3.99	<0.001	1925–2001
Fringilla coelebs	79		0.127	574.6	0.263	−0.07	4.06	0.561	1927–2002
Carduelis chloris	52	X	−0.67	311.9	<0.001	0.605	3.51	<0.001	1922–1980
Carduelis carduelis	33	X	0.371	664.8	0.034	−0.27	5.47	0.132	1928–1998
Carduelis spinus	60		−0.09	522.1	0.518	0.272	4.84	0.035	1959–1984
Pyrrhula pyrrhula	58		−0.28	223.6	0.033	0.206	2.07	0.12	1954–1976
Coccothraustes coccothraustes	111		−0.10	440.5	0.299	0.223	3.24	0.019	1954–1985
Emberiza citrinella	31		−0.35	373.2	0.056	0.241	3.35	0.192	1929–1977
Emberiza schoeniclus	40		0.060	481.1	0.714	−0.04	3.45	0.805	1962–1992

[a] After restrictions as described in text.

[b] Evidence for increased numbers of winter recoveries at a distance below 100 km.

[c] Mean recovery distance [decrease negative, increase positive; β is the regression coefficient (year of recovery/recovery distance); SE the standard error of estimate].

[d] Mean latitude of winter recoveries [further south: negative, further north: positive; β is the regression coefficient (year of recovery/recovery latitude); SE the standard error of estimate].

[e] Period regarded.

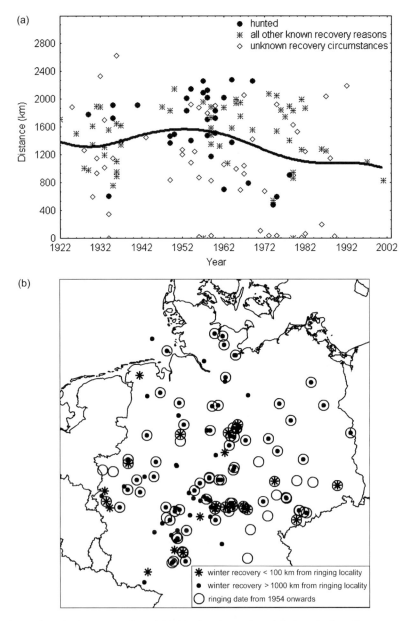

Figure 2 Ringing recoveries of the black redstart ringed during the breeding season in Germany and recovered November–February. (a) Recovery distances (km) against years. The line is the least squares fit. (b) Ringing localities of birds ringed during the breeding season and recovered in winter.

assignment of newly emerging winterers to distinct populations deserves priority in future ringing projects because it allows comparison of their behaviour with that of migratory populations to test hypotheses about the evolution of bird migration.

C. Changes in Recovery Distances

Recovery distances are the calculated lengths of loxodromes between the coordinates of ringing and those of recovery (Imboden and Imboden, 1972). Mean recovery distances refer to the mean of all distances present in a dataset for a defined time-span.

From reported field observations we can state the hypothesis that recovery distances in some species should decrease as birds start to winter closer to their breeding areas, when environmental conditions allow local wintering, while it should increase if birds extend their migration distance. Among the 30 tested species we found evidence for reduced mean recovery distances with time in eight species and evidence for increased mean recovery distances in five species (Table 1). Again, this is more than the 1.5 species that are expected to show the change in the predicted direction with a significance level of 5%. Figure 3 shows two examples for decreased recovery distances which are likely to represent decreases in migration distances with time (lapwing *Vanellus vanellus*, linear regression 1950–2001: $\beta = -0.27$, $R^2 = 0.07$, $p < 0.001$ and greenfinch *Carduelis chloris*, linear regression 1922–1980: $\beta = -0.67$, $R^2 = 0.45$, $p < 0.001$).

While the results for these eight species meet the expectation that under warmer climate conditions birds are likely to start wintering closer to their breeding areas, the five species with opposite results demonstrate the weakness of this approach. However, some of the detected increases in mean recovery distances can be attributed to methodological problems. Data for the wood pigeon *Columba palumbus* (linear regression 1922–1999: $\beta = 0.35$, $R^2 = 0.12$, $p < 0.001$) demonstrate the strong influence of hunting habits on temporal patterns of recovery (Figure 4). The decrease in number of pigeons reported as hunted in Germany since the late 1970s lead to a lack of short-distance recoveries during winter. In contrast, numbers of winter sightings of Wood Pigeons in Germany did not decrease (Bauer *et al.*, 1995). The same might apply to the stock dove *Columba oenas*, but sample size does not allow any tests. In the lapwing (Figure 3) most of the decrease in recovery distances could also be explained by an absence of recoveries of hunted birds at distances exceeding 1300 km since 1985. However, there is no evidence for this alternative explanation in the areas concerned (Southern France, Northern Spain), since reported recoveries due to hunting for many other species are still common for this area.

(a)

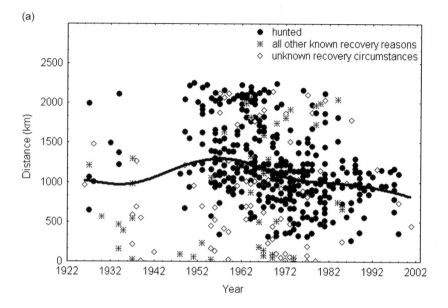

(b)

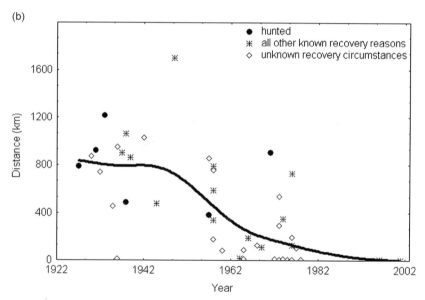

Figure 3   Recovery distances (km) between November and February of lapwings (a) and greenfinches (b) ringed during the breeding season in Germany.

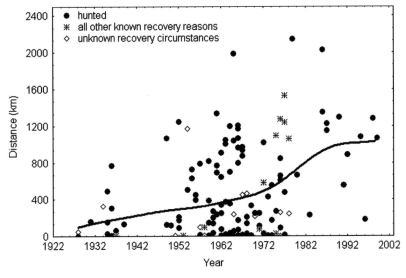

Figure 4   Recovery distances between November and February of wood pigeons ringed during the breeding season in Germany.

In the goldfinch *Carduelis carduelis*, starling *Sturnus vulgaris* (Figure 5; linear regression 1922–2001: $\beta = 0.57$, $R^2 = 0.33$, $p < 0.001$) and white wagtail *Motacilla alba* methodical reasons for the observed pattern of increasing recovery distances are not obvious. The same applies to the grey heron *Ardea cinerea* (Figure 5; linear regression 1925–2000: $\beta = 0.06$, $R^2 = 0.003$, n.s.) where no significant trend in recovery distances could be found. For these species a convincing explanation is still absent [for example increasing wintering numbers of grey herons in South-western Germany are reported from field counts, e.g., Bauer *et al.* (1995)]. Future studies should investigate effects of population density on mean recovery distance. It is possible that if the population of a species is increasing, we might still see a stable or even an increasing fraction migrating long distances. An increase in migratory tendency with population density was shown for English greenfinches by Main (1996), and Soutullo (2003) found a positive correlation between species with positive population trends in Britain and Ireland and their tendency to winter further south.

Reduced mean recovery distances with time were recently reported in a number of studies (e.g., Cooke *et al.* (1995) for North American snow geese *Anser caerulescens*, Bairlein (1992) and Fiedler (2001) for European white storks *Ciconia ciconia*, Švažas *et al.* (2001) for Lithuanian mute swans *Cygnus olor*, mallards *Anas platyrhynchos* and coots *Fulica atra*, Remisiewicz (2001) for Polish mallards, Siefke (1994) for European crows *Corvus corone cornix*).

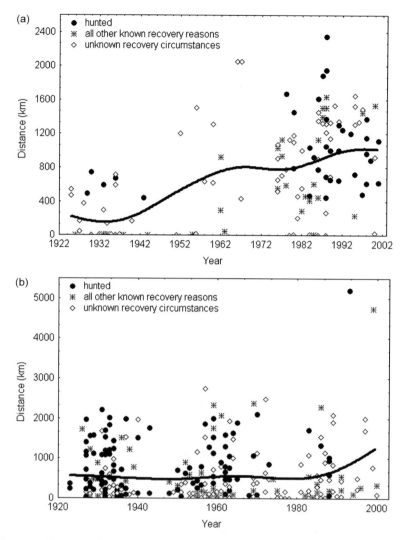

Figure 5   Recovery distances between November and February of starlings (A) and grey herons (B) ringed during the breeding season in Germany.

The most elaborate methods to investigate changes in migration as revealed by recovery distances have been proposed by Siriwardena and Wernham (2002) who compared a calculated index of migratory tendency of British and Irish breeding bird species before and after the median year of all recoveries available. This method is especially suitable to detect smaller changes in partial migrants which form the group of birds in which we expect most changes to occur.

D. Changes in Recovery Latitude

Analysing only the mean latitudes of recoveries of birds ringed during the breeding season within a defined area and recovered during winter reduces the risk of introducing biases through variation in regional ringing efforts in long-term analyses. Table 1 shows the results for the 30 species from the German ringing recovery dataset. A tendency towards wintering at higher latitudes was found in 10 species while mean latitudes of winter recoveries moved southwards in three species. These results did not contradict with those derived from the analysis of recovery distances. Soutullo (2003) found that between 1960 and 1997 the average wintering latitude of 27 out of 66 bird species of Great Britain and Ireland showed wintering to occur further north (Table 2).

E. Changes in Recovery Directions

Recovery directions (compass direction of the recovery coordinates as seen from the site of ringing) were analysed as described above. In 28 of 30 species no evidence for changes of mean directions with time was found. The two remaining species were coot, where changes in the focal areas of ringing resulted in a change of recovery directions, and common gull *Larus canus*. While the German data for the common gull did not show any trends in mean recovery distances, breeding birds from the German Baltic coast showed an increasing winter presence in the Wadden Sea as compared to the Belgian and French Atlantic coast. Consequently, this caused a 40–45° northward shift of migration directions in recovery distances above 200 km.

Blackcaps (*Sylvia atricapilla*) from southern Germany and neighbouring areas started wintering in Great Britain and even Southern Sweden [shift in recovery directions from south-westerly to north-westerly and northern directions (Berthold and Terrill, 1988; Fransson and Stolt, 1993)]. Changes in recovery directions were also found for Central European white storks (Bairlein, 1992; Fiedler, 2001), where birds from the Netherlands and parts of central Europe shifted from south-easterly to south-westerly migratory directions. However, these populations received a considerable influx of trans-located or otherwise manipulated individuals during the so-called reintroduction projects. This is the most likely reason for a change in migration direction.

F. Comparison of Results and Methods

A comparative overview of our results based on German ringing recoveries and those of the British Isles is given in Table 2. A few inconsistencies within the same area still need to be explained, e.g., in the case of British and Irish lapwings

Table 2 Comparison between results of the studies of the German and British/Irish ringing recovery data. Indications in brackets are based on $p < 0.1$, all other results are based on $p < 0.05$. Plus indicates an increase, minus a decrease, O not significant results; fields were left blank when test was not performed

Species	German data (this study)				Siriwardena and Wernham (2002)		Soutullo (2003)	
	Period[a]	Winter recoveries[a]	Latitude[a]	Distance[a]	Median[b]	Migration tendency[c]	Latitude (1960–1997)	Latitude (1909–1959)[d]
Phalacrocorax carbo	1965–1996	+	O	O	1973	O	O	O
Ardea cinerea	1925–2000	O	O	O	1972	O	+	O
Tadorna tadorna	1950–1973	+	+	–	Mid 1974	O	+	–
Anas plathyrhynchos	1929–2000	O	O	O	1970	O	O	
Buteo buteo	1660–1986	+	+	O	1982	–	O	O
Accipiter nisus	1922–2002	+	+	–	1985	O	–	O
Falco tinnunculus	1928–2001	O	O	(–)	1982	–	–	O
Fulica atra	1954–1991	+	O	(–)	Mid 1973	O		
Haematopus ostralegus	1927–2001	O	O	O	1977	–	+	O
Charadrius hiaticula	1953–1986	O	(+)	(–)	Mid 1967	O		
Vanellus vanellus	1950–2001	O	+	–	1953	+	+	–
Larus ridibundus	1928–2002	O	+	(–)	1966	–	O	O
Larus canus	1928–1994	O	O	–	1983	O	O	O

Species[a]	[b,c]	[b,c]	Period[d]	[b,c]	[b,c]	Mid 1970	[b,c]	[b,c]
Columba oenas	O	O	1978–2001	O	+		O	O
Columba palumbus	O	—	1922–1999	—	+	1967	—	—
Motacilla alba	O	—	1950–1992	—	O	1968	+	+
Prunella modularis	O	O	1934–1979	O	O	1975	O	O
Turdus merula	O	+	1995–2002	+	—	1971	+	O
Turdus philomelos	O	O	1924–2000	O	+	1964	+	O
Sturnus vulgaris	O	—	1925–2001	—	O	1965	(+)	—
Fringilla coelebs	O	O	1927–2002	O	+	1977	+	—
Carduelis chloris	+	+	1922–1980	+	—	1975	(+)	—
Carduelis carduelis	+	O	1928–1998	O	+	1971	+	
Pyrrhula pyrrhula	O	O	1954–1976	O	—	1973	O	
Emberiza schoeniclus	O	O	1962–1992	O	O	1979	O	
Number of species tested[e]	30	9		73	15		66	38
Evidence for wintering closer to breeding grounds		10 (11)	8 (13)			27 (30)	3 (3)	
Evidence for wintering farther away		3 (3)	5 (5)	5		11 (11)	11 (11)	
No trend detected/more complex changes		17 (16)	17 (12)	51/2		28 (25)	24 (24)	

[a] See Table 1.
[b] Changes are derived by comparison of the migratory tendency on both sides of this median year available in the dataset.
[c] Migration tendency (see text).
[d] Relationships between wintering latitude and year are calculated separately for the time periods before and since 1960.
[e] Including those not tested in both Germany and Britain/Ireland and, therefore, not included in the upper part of this table.

which show an increasing migration tendency according to Siriwardena and Wernham (2002), but at the same time the mean winter recovery latitude moves northward according to Soutullo (2003). Five out of the 25 species analysed in both areas show evidence for a reduction in migratory activity and two species (wood pigeon and starling) show evidence for the opposite trend. Obviously different trends between areas can only be seen for sparrowhawk *Accipiter nisus*, white/pied wagtail and song thrush *Turdus philomelos*.

A conclusion about the different measures of change and their advantages and disadvantages are given in Table 3.

IV. IDENTIFICATION OF UNDERLYING FACTORS AND THE ROLE OF CLIMATE CHANGE

So far we only presented a description of changes in the migratory behaviour based on the ringing recovery data. For most of these, an interpretation related to global warming and especially increasing temperatures in Central Europe seems plausible. However, a serious scientific test of whether climate change is an important factor responsible for the observed changes is still lacking. This applies not only to the data presented here, but also to most other recently reported changes in bird migration behaviour based on the ringing data. Siriwardena and Wernham (2002) tested individuals of 73 bird species ringed in Britain and Ireland, and they found a decrease in migratory tendency in 15 and an increase in five species. However, they state that this pattern would be predicted if global warming had increased the suitability of more northerly regions for wintering birds. Soutullo (2003) showed a relationship between wintering latitude and changes in temperature, rainfall and the North Atlantic Oscillation Index (NAO, Hurrell *et al.*, 2001). However, these patterns are not straightforward, since the directions of relationship between wintering latitude and climate variables were different in different species (Soutullo, 2003). This suggests that species-specific attributes related to ecology or population density have to be invoked to explain the observed patterns.

Stochasticity in the ringing recovery data may conceal relationship between climate factors and responses as revealed by the recovery data. In addition, selecting relevant climate data is important before embarking on appropriate statistical tests. A large-scale index to describe precipitation and temperature in summer and autumn in continental Europe is not available. NAO seems to be suitable in Atlantic regions and might be used to describe inland situations as well, although geographical variation is large (Hurrell *et al.*, 2001). Changes in positions of wintering areas as revealed by ring recoveries are likely to be influenced by many factors other than climate (Fiedler, 2003). At the same time many of these changes do not contradict expectations from climate data such as

Table 3 Comparison of methods to detect and measure changes in migratoryness of bird populations using ringing recovery data

Measure of change	Brief description	Advantages	Problems
Local winter recoveries	Check presence of local winter recoveries, e.g., within a distance less than 100 km around the place of ringing (which happened during the breeding season)	In combination with field observations particularly strong evidence for changes in migration pattern. Additional value to field observations due to knowledge about the origin of the observed winterers	Pattern might be influenced by: (1) changes in causes of death (such as hunting!), (2) changes in ringing activity, (3) changes in recapture and re-sighting activity and (4) changes in population size over time
Recovery distances	Calculate relationship between time and mean distances between place of ringing (during breeding season) and place of recovery (during winter season)	Gives a much more detailed and continuous figure than the method above. Also detects trends towards a higher migratory activity	As above. A few long distance recoveries might occur or be missed randomly and thus mask changes on small scales
Recovery latitude	Calculate relationship between time and mean latitude of recovery of all individuals ringed in a given area and recovered during winter season	Reduces the risk of introducing biases through variation in regional ringing activities	Does not detect changes in west–east (e.g., Atlantic–continental) direction. Sensitive to changes in recovery probability and population size
Migration tendency	Calculate migration tendency from the shape of the frequency distributions of distances between ringing during breeding season and recovery during winter season [two approaches, see Siriwardena and Wernham (2002)]. Compare results from time periods (e.g., at both sides of the median year of all recoveries)	Especially suitable to detect smaller changes in partial migrants. A way to quantify migratory strategy (sedentary, partially migratory to different degrees, migratory)	Computations more elaborate. Pattern might be influenced by same factors as indicated in local winter recoveries above

an increasing mean temperature or drier winters with less snow resulting in better foraging conditions in certain areas. For instance, changes in wintering areas linked to changes in human land use are reported for the European crane *Grus grus* (Alonso *et al.*, 1994) and the red-breasted goose *Branta ruficollis* (Vangeluwe and Stassin, 1991). Other human activities like winter-feeding by humans have improved the conditions for continental blackcaps during winter in Great Britain (Berthold and Terill, 1988). Likewise, the availability of open garbage-dumps may have caused the local wintering behaviour of Danish herring gulls to change to wintering closer to their breeding grounds (Petersen, 1984). Similarly, ice-free waters in reservoirs next to power plants may have allowed coots to stay near their breeding grounds in Lithuania instead of migrating more than 1000 km south-westwards (Švažas *et al.*, 2001). Even changes in hunting pressure may effect the establishment of new wintering areas as shown for the Eurasian wigeon *Anas penelope* in Denmark (Madsen, 1985).

These and many more examples may provide sufficient caution not to attribute detected changes in migration behaviour unconditionally to climate change until convincing causalities have been investigated. We expect that many changes in bird migration recently reported from field observations and ringing studies are indeed induced by climate change. However, we regard the rigorous tests to be one of the most demanding challenges within this field of avian science. Large-scale analysis as we performed for the German ringing and recovery data is suitable for searching large databases and pinpointing species, regions and time-spans where much more detailed investigations may prove to be promising.

REFERENCES

Adriaensen, F., Ulenaers, P. and Dhondt, A.A. (1993) *Ardea* **81**, 59–70.
Alonso, J.C., Alonso, J.A. and Bautista, L.M. (1994) *J. Appl. Ecol.* **31**, 212–222.
Bairlein, F. (1992) In: *Les Cigogne d'Europe* (Ed. by Institut Europeen d'Ecologie and A.M.B.E.), pp. 191–205, Institut d'Ecologie, Metz.
Bauer, H.-G., Boschert, M. and Hölzinger, J. (1995) In: Die Vögel Baden Württembergs (Ed. by J. Hölzinger), Vol. 5, Ulmer, Stuttgart.
Berthold, P. and Terrill, S.B. (1988) *Ringing and Migration* **9**, 153–159.
Cooke, F., Rockwell, R.F. and Lank, D.B. (1995) *The Snow Gees of La Perouse Bay.* Oxford University Press, Oxford.
Dagys, M. (2001). In: *Changes of Wintering Sites of Waterfowl in Central and Eastern Europe* (Ed. by S. Švažas, W. Meissner, V. Serebryakov, A. Kozulin and G. Grishanov), pp. 24–28. OMPO, Vilnius.
Fiedler, W. (2001). *Ring* **23**, 73–79.
Fiedler, W. (2003) In: *Avian Migration* (Ed. by P. Berthold, E. Gwinner and E. Sonnenschein), pp. 21–38. Springer, Berlin.

Fiedler, G. and Wissner, A. (1989) In: *White Stork. Proceedings I. International Stork Conservation Symposium* (Ed. by G. Rheinwald, J. Ogden and H. Schulz), pp. 423–424. DDA, Berlin.

Fransson, T. and Stolt, B.-O. (1993) *Vogelwarte* **37**, 89–95.

George, K. (1995). *Vogelwelt* **116**, 311–315.

Glutz von Blotzheim, U. (Ed.) (1966–1997) *Handbuch der Vögel Mitteleuropas.* Aula, Wiesbaden.

Hurrell, J.W., Kushnir, Y. and Visbeck, M. (2001) *Science* **291**, 603–605.

Imboden, C. and Imboden, D. (1972) *Vogelwarte* **26**, 336–346.

Loonen, M. and De Vries, C. (1995) *Limosa* **68**, 11–14.

Madsen, J. (1985). *Bird Study* **127(suppl. 1)**, 67–74.

Main, I. (1996) *Bird Study* **43**, 240–252.

Møller, A.P. (1981) *Vogelwarte* **31**, 74–94, see also pp. 149–168.

Petersen, B.S. (1984) *Dansk Ornithol. Foren. Tidsskr.* **78**, 15–24.

Remisiewicz, M. (2001) In: *Changes of Wintering Sites of Waterfowl in Central and Eastern Europe* (Ed. by S. Švažas, W. Meissner, V. Serebryakov, A. Kozulin and G. Grishanov), pp. 82–93. OMPO, Vilnius.

Siefke, A. (1994) *Vogelwelt* **115**, 83–89.

Siriwardena, G.M. and Wernham, C.V. (2002) In: *The Migration Atlas: Movements of the Birds of Britain and Ireland* (Ed. by C.V. Wernham, M.P. Toms, J.H. Marchant, J.A. Clark, G.M. Siriwardena and S.R. Baillie), T.& A.D. Poyser, London.

Soutullo, A. (2003) Dissertation University of East Anglia, Norwich (cited with friendly permission of the author).

Speek, G., Clark, J.A., Rohde, Z., Wassenaar, R.D. and van Noordwijk, A.J. (2001) The EURING exchange-code 2000, Heteren.

Švažas, S., Patapavičius, R. and Dagys, M. (2001) In: *Changes of Wintering Sites of Waterfowl in Central and Eastern Europe* (Ed. by S. Švažas, W. Meissner, V. Serebryakov, A. Kozulin and G. Grishanov), pp. 56–64. OMPO, Vilnius.

Vangeluwe, D. and Stassin, P. (1991) *Gerfaut* **81**, 65–99.

Wernham, C.V. and Siriwardena, G.M. (2002) In: *The Migration Atlas: Movements of the Birds of Britain and Ireland* (Ed. by C.V. Wernham, M.P. Toms, J.H. Marchant, J.A. Clark, G.M. Siriwardena and S.R. Baillie), T.& A.D. Poyser, London.

Breeding Dates and Reproductive Performance

PETER DUNN*

I. SUMMARY

Much of the evidence for the effects of climate change on birds comes from studies of the timing of breeding. Many species of birds start to lay eggs earlier in years with warmer temperature, and approximately 60% of studies have shown long-term advances in laying dates consistent with global warming. Nevertheless, the magnitude of the responses differs among species and locations in ways that we do not yet understand. Some of the variation is probably due to differences in: (1) local temperature changes, (2) diet, (3) body size, (4) life history (migratory or resident; number of broods per season) and (5) the time scales over which species acquire resources for breeding. Although earlier laying of clutches is often associated with larger clutch sizes and greater production of young, the effects of earlier laying on reproductive performance are less clear. In the future, we need to establish large-scale collaborative studies to monitor the effects of climate change over large areas and to determine how climate change affects the reproductive performance of species throughout their ranges.

E-mail address: pdunn@uwm.edu (P. Dunn)

ADVANCES IN ECOLOGICAL RESEARCH, VOL. 35
0065-2504/04 $35.00 DOI 10.1016/S0065-2504(04)35004-X

II. INTRODUCTION

Timing of breeding is one of the most important factors influencing reproductive performance in birds. Individuals that breed early tend to produce larger clutches and more surviving offspring than individuals that breed later (Lack, 1968; Price and Liou, 1989). In many species, photoperiod is the primary cue to begin gonadal development and courtship, whereas ambient temperature and food availability are used for finer scale decisions about the timing of breeding (Hahn et al., 1997; Pulido and Berthold, 2004, this volume). Females appear to time the laying of their clutch such that hatching occurs near the peak of food availability when the energetic demands of breeding are presumably the greatest. However, food availability during egg-laying may be important in some species (Perrins, 1970; Bryant, 1975), and other aspects of life history, such as the number of broods per season (Crick et al., 1993; Visser et al., 2003), can influence the optimal time to lay eggs. In many bird species, the timing of breeding is synchronised with peaks of insect abundance (reviewed by Daan et al., 1988) that are often correlated with temperature (Visser and Holleman, 2001). Spring temperatures are rising and are expected to increase at a higher rate during the remainder of the 21st century (Houghton et al., 2001). Although many bird species appear to be advancing their breeding in response to climate change, the trends vary among species and region.

To predict the effects of climate change on timing of breeding we need to know: (1) how much temperature is changing in particular locations, and (2) how temperature affects timing of breeding and subsequent reproductive performance. In this review, I examine the state of our current knowledge of how temperature, and climate change in particular, affect reproductive performance in different species and locations.

III. CHANGES IN TEMPERATURE

On a global scale, temperatures have increased by 0.6°C over the past century (Houghton et al., 2001); however, some regions are warming more and others less than the global average (see Figure 1 in Walthers et al., 2002). For example, in the northern hemisphere, the greatest increases in temperature and photosynthetic activity have occurred above 50°N, although some parts of the midwestern United States and eastern Asia have also experienced strong increases (Myneni et al., 1997; Walthers et al., 2002). These regional differences in temperature change are likely to affect migratory and resident species differently. Migratory species are more likely to time their arrival on the breeding grounds based on photoperiod rather than local temperature, and, if spring temperatures continue to increase, migratory species may arrive too late to take full advantage of changes in food availability. For example, over the past 20 years there has been an increase in average spring temperature in the Netherlands, which has led to an advancement in the peak of insect abundance. Nevertheless, pied flycatchers (*Ficedula hypoleuca*) breeding in the Netherlands continue to arrive from their wintering grounds in Africa at about the same time each year (Both and Visser, 2001). The interval between arrival and egg-laying has decreased, but this advancement has not kept up with changes in the peak of food abundance (Both and Visser, 2001). As a consequence, pied flycatchers are breeding later relative to the peak in food abundance. In contrast, resident birds are present year-round and can potentially track any changes in the food supply that occur before migratory species arrive. Of course, resident species may have their own constraints (spring moult or severe spring weather) that limit their ability to take advantage of these early changes in food supply. Overall, however, the regional differences in temperature change combined with variation in the movement patterns of birds lead to a potentially complex assortment of temperature effects in different species (see also Sæther *et al.*, 2003).

IV. HOW DOES TEMPERATURE AFFECT THE TIMING OF REPRODUCTION?

Numerous studies have found that most north temperate birds start to lay their eggs earlier in the season when spring temperatures are higher. In a survey of the literature, 79% (45/57) of species showed a significant negative relationship between the date of laying and air temperature (Table 1). In collared flycatchers (*F. albicollis*) timing of laying is also related negatively to the North Atlantic Oscillation (NAO) index (Przybylo *et al.*, 2000), which is associated positively with warmer and wetter winters in coastal Europe (Hurrell, 1995). There are several ways in which the correlation between laying date and temperature may arise. First, there may be a direct effect of temperature on the energetic demands of females which influences their timing of laying. In a study of captive starlings (*Sturnus vulgaris*) with *ad libitum* food, Meijer *et al.* (1999) heated and cooled nest-boxes by $2-3°C$ and found that temperature had a direct effect on the timing of laying independent of food supply and photoperiod. Second, temperature may influence the growth of gonads, which could indirectly affect the timing of laying. For example, the testes of black-billed magpies (*Pica pica*) kept in the laboratory exhibited faster growth under long day conditions at 20°C than at 2°C (e.g., Jones, 1986). Precipitation could have a similar effect, as the growth of spotted antbird (*Hylophylax naevioides*) testes and follicles in Panama was slower in a dry year (Wikelski *et al.*, 2000). In any case, there are likely several direct and indirect relationships between food, temperature and gonadal development. Lastly, temperature may influence food availability, particularly

Table 1 Studies of the relationship between laying date and air temperature

Species	Location	Slope	Significance	n nests	Years	Reference
American coot (*Fulica americana*)	North America		ns	216	1956–1983	Torti and Dunn (unpubl. data)
American robin (*Turdus americana*)	North America	−0.71	*	2695	1951–1999	Torti and Dunn (unpubl. data)
Barn swallow (*Hirundo rustica*)	Great Britain	−2.76	*	2893	1962–1995	Crick and Sparks (1999)
Black-billed magpie (*Pica pica*)	Great Britain		ns	1326	1942–1995	Crick and Sparks (1999)
Blackcap (*Sylvia atricapilla*)	Great Britain	−5.37	*	1397	1940–1995	Crick and Sparks (1999)
Blue tit (*Parus caeruleus*)	France		*		1979–1989	Bellot et al. (1991)
Bullfinch (*Pyrrhula pyrrhula*)	Great Britain	−6.98	*	1689	1939–1995	Crick and Sparks (1999)
Carrion crow (*Corvus corone*)	Great Britain		ns	1258	1939–1995	Crick and Sparks (1999)
Chaffinch (*Fringilla coelebs*)	Great Britain	−4.42	*	4651	1939–1995	Crick and Sparks (1999)
Chiffchaff (*Phylloscopus collybita*)	Great Britain	−3.98	*	1255	1937–1995	Crick and Sparks (1999)
Cliff swallow (*Petrochelidon pyrrhonota*)	USA		*	1093	1982–1989	Brown and Brown (1999)
Collared dove (*Streptopelia decaocto*)	Great Britain		ns	1088	1960–1995	Crick and Sparks (1999)
Common moorhen (*Gallinula chloropus*)	Great Britain		*	2399	1962–1995	Crick and Sparks (1999)
Common redstart (*Phoenicurus phoenicurus*)	Great Britain	−1.11	ns	2078	1941–1995	Crick and Sparks (1999)
Common coot (*Fulica atra*)	Netherlands		*	204		Perdeck and Cave (1989)
Dunnock (*Prunella modularis*)	Great Britain	−2.33	*	4106	1939–1995	Crick and Sparks (1999)
Common crane (*Grus grus*)	Ukraine	−1.19	*	100		Winter et al. (1999)
Eastern bluebird (*Sialis sialis*)	North America	−0.62	*	9494	1950–2000	Torti and Dunn (unpubl. data)
Eurasian oystercatcher (*Haematopus ostralegus*)	Great Britain		ns	1533	1962–1995	Crick and Sparks (1999)
European robin (*Erithacus rubecula*)	Great Britain	−3.51	*	5909	1939–1995	Crick and Sparks (1999)
European starling (*Sturnus vulgaris*)	Great Britain	−3.27	*	3838	1940–1995	Crick and Sparks (1999)
European starling	Germany	−0.24	*	164	1981–1990	Meijer et al. (1999)

Species	Country	Value	Sig	N	Years	Reference
Great tit (*Parus major*)	Switzerland	−3.84	*		1988–1992	Nager and van Noordwijk (1995)
Great tit	Great Britain	−0.08	*		1947–1997	McCleery and Perrins (1998)
Great tit	Netherlands		*		1973–1995	Visser *et al.* (1998)
Great tit	Finland		*	93	1983–1994	Eeva *et al.* (2000)
Greenfinch (*Carduelis chloris*)	Great Britain	−3.73	*	3457	1939–1995	Crick and Sparks (1999)
Grey wagtail (*Motacilla cinerea*)	Great Britain		ns	2118	1949–1995	Crick and Sparks (1999)
Hawfinch	Netherlands	−3.50	*	249	1989–1997	Bijlsma (1998)
(*Coccothraustes coccothraustes*)						
Killdeer (*Charadrius vociferus*)	North America	−2.08	*	319	1961–1997	Torti and Dunn (unpubl. data)
Linnet (*Carduelis cannabina*)	Great Britain	−2.92	*	4269	1939–1995	Crick and Sparks (1999)
Little swift (*Apus affinis*)	Japan		*	96	1986–1989	Hotta (1996)
Long-tailed tit (*Aegithalos caudatus*)	Great Britain	−5.51	*	1420	1940–1995	Crick and Sparks (1999)
Marsh harrier (*Circus aeruginosus*)	Netherlands	−0.57	*	1379	1975–1994	Dijkstra and Zijlstra (1997)
Meadow pipit (*Anthus pratensis*)	Great Britain	−3.93	*	1656	1950–1995	Crick and Sparks (1999)
Mexican jay (*Aphelocoma ultramarina*)	USA	−0.40	*	458	1971–1998	Brown *et al.* (1999)
Mistle thrush (*Turdus viscivorus*)	Great Britain	−5.21	*	1385	1939–1995	Crick and Sparks (1999)
Northern lapwing (*Vanellus vanellus*)	Great Britain		ns	1078	1962–1995	Crick and Sparks (1999)
Pied flycatcher (*Ficedula hypoleuca*)	Netherlands	−1.67	*	414	1980–2000	Both and Visser (2001)
Pied flycatcher	Finland		*	470	1983–1994	Eeva *et al.* (2000)
Pied wagtail (*Motacilla alba*)	Great Britain	−0.75	*	2925	1939–1995	Crick and Sparks (1999)
Redstart (*P. phoenicurus*)	Finland		*	138	1983–1994	Eeva *et al.* (2000)
Red-winged blackbird	North America	−1.02	*	3868	1956–2000	Torti and Dunn (unpubl. data)
(*Agelaius phoeniceus*)						
Reed bunting (*Emberiza schoeniclus*)	Great Britain	−2.64	*	2248	1943–1995	Crick and Sparks (1999)
Reed warbler (*Acrocephalus scirpaceus*)	Great Britain	−1.24	ns	5305	1940–1995	Crick and Sparks (1999)
Ringed plover (*Charadrius hiaticula*)	Great Britain	−1.09	*	1332	1944–1995	Crick and Sparks (1999)

(Continued)

Table 1 Continued

Species	Location	Slope	Significance	n nests	Years	Reference
Sedge warbler (*Aerocephalus schoenobaenus*)	Great Britain	−2.04	*	2218	1940–1995	Crick and Sparks (1999)
Siberian tit (*Poecile cinctus*)	Finland		ns	113	1983–1994	Eeva et al. (2000)
Song sparrow (*Melospiza melodia*)	North America		*			Arcese et al. (2002)
Song thrush (*Turdus philomelos*)	Great Britain	−1.23	*	6388	1962–1995	Crick and Sparks (1999)
Spotted flycatcher (*Muscicapa striata*)	Great Britain	−2.72	*	3294	1939–1995	Crick and Sparks (1999)
Tree sparrow (*Passer montanus*)	Great Britain	−4.83	*	4034	1942–1995	Crick and Sparks (1999)
Tree swallow (*Tachycineta bicolor*)	North America		*	3450	1959–1991	Dunn and Winkler, (1999)
Whinchat (*Saxicola rubetra*)	Great Britain		ns	1066	1941–1995	Crick and Sparks (1999)
White-backed woodpecker (*Dendrocopus leucotos*)	Norway		*	69	1983–1994	Hogstad and Stenberg (1997)
White-tailed ptarmigan (*Lagopus leucurus*)	USA	−2.32[a]	*		1975–1999	Wang et al. (2002)
Whitethroat (*Sylvia communis*)	Great Britain		ns	1108	1940–1995	Crick and Sparks (1999)
White-throated dipper (*Cinclus cinclus*)	Great Britain	−1.59	*	2074	1943–1995	Crick and Sparks (1999)
Willow warbler (*Phylloscopus trochilus*)	Great Britain	−1.47	*	3931	1939–1995	Crick and Sparks (1999)
Winter wren (*Troglodytes troglodytes*)	Great Britain	−3.88	*	3620	1939–1995	Crick and Sparks (1999)
Wood warbler (*Phylloscopus sibilatrix*)	Great Britain	−1.93	*	1301	1941–1995	Crick and Sparks (1999)
Woodcock (*Scolopax rusticola*)	Great Britain		*	449	1945–1989	Hoodless and Coulson (1998)
Yellowhammer (*Emberiza citrinella*)	Great Britain		ns	1181	1939–1995	Crick and Sparks (1999)

Blanks indicate data were not reported. Slope is from linear regressions. Note that studies of the relationship between laying date and other climate variables (NAO, SOI) were not included. * indicates $p < 0.05$ and ns indicates $p > 0.05$ from linear regressions or Spearman correlations. Relationships were all negative when significant.

[a] Hatch date was analysed rather than laying date.

insects (e.g., Bryant, 1975), which, in turn, may limit the ability of females to produce eggs (Perrins, 1970). These mechanisms are not exclusive and are likely to act together to influence timing of laying.

V. CHANGES IN LAYING DATE AND REPRODUCTIVE PERFORMANCE

Collections of nest records, such as those at the British Trust for Ornithology and the Cornell Laboratory of Ornithology, have provided crucial resources for studying the effects of temperature on reproductive performance. Recent studies based on nest records have provided mounting evidence that birds are starting to lay clutches earlier in response to warmer temperatures (Crick and Sparks, 1999; Dunn and Winkler, 1999). In the United Kingdom, 31% (20/67) of bird species showed significant trends towards earlier laying over the 25-year period from 1971 to 1995 (Crick et al., 1997). On average, eggs were laid 8.8 days earlier in 1995 than in 1971 (Crick et al., 1997). In a subsequent study, Crick and Sparks (1999) reported that 19 of 36 (53%) species in the UK are now breeding earlier than they were in the 1960s and 1970s, and of these 19 species, 17 also showed a negative correlation between laying date and spring (March and April) temperature or precipitation. Advances in the date of laying or hatching of eggs have also been reported in long-term studies at particular locations in 27 species (Table 2). Note, however, that in some well-studied species, such as pied flycatchers, great tits (Parus major) and blue tits (P. caeruleus), there is considerable variation among study sites (e.g., Visser et al., 2003; see also Sæther et al., 2003). To date, the only long-term, large-scale studies of the effects of climate change on the timing of breeding of birds have been conducted on tree swallows (Tachycineta bicolor) in North America (Dunn and Winkler, 1999) and pied flycatchers (Sanz, 2003) and great and blue tits in Europe (Sanz, 2002; Visser et al., 2003; Figure 1). These studies reveal an overall pattern of advancing laying date where temperatures (or the NAO) have increased, but there are also significant regional differences that we do not yet understand.

We might expect warmer temperatures to lead to greater production of young, because laying earlier is often associated with larger clutch size and more young fledged (Lack, 1968). However, recent evidence from great tits suggests that warmer spring temperature can lead to a mismatch in the timing of egg-laying relative to the availability of food for nestlings, and, as a consequence, earlier laying females may produce fewer surviving young (Visser et al., 1998; Coppack and Pulido, 2004, this volume). Thus, it is important to examine the effects of climate change on all aspects of reproduction, not just laying date. In particular, changes in clutch size are likely to have a major impact on fitness, as it places an upper limit on total reproductive success for a given brood. Long-term increases

Table 2 Studies of effects of climate or temperature change on breeding parameters of birds

Species	Climate variable	Has climate changed?	Breeding parameter	Relationship between breeding parameter and		Reference
				Climate variable	Year	
36 species in UK	Temp.	Yes, +	Laying date	− (58% of spp.)	Earlier (53% of spp.)	Crick and Sparks (1999)
White-tailed ptarmigan (*Lagopus leucurus*)	Temp.	Yes, +	Hatch date	−	Earlier	Wang et al. (2002)
Capercaillie (*Tetrao urogallus*)	Temp.	Variable[a]	Chicks per hen	+	Declining	Moss et al. (2001)
			% hens with chicks	+	Declining	
Tufted puffin (*Fratercula cirrhata*)	SST[b]	Yes, +	Hatch date	−	Earlier	Gjerdrum et al. (2003)
Lesser kestrel (*Falco naumanni*)	Rainfall	Yes, − (P = 0.056)	Fledging success	Curvilinear		Rodriguez and Bustamente (2003)
			% occupied nest cavities	Curvilinear		
			Female chicks per breeding female	−		
Emperor penguin (*Aptenodytes forsteri*)	Extent of pack ice		Fledging success	−		Barbraud and Weimerskirch (2001)
Northern fulmar (*Fulmaris glacialis*)	NAO[c]	Yes, +	Hatching and fledging success	−		Thompson and Ollason, (2001)
Mexican jay (*Aphelocoma ultramarina*)	Temp.	Yes, +	Laying date	−	Earlier	Brown et al. (1999)
Pied flycatcher (*Ficedula hypoleuca*)	Temp.	Yes, +	Laying date	−	Earlier	Järvinen (1989)
			Clutch size		Increasing	
			Fledging success		0	
		Yes, +	Hatching date	−	Earlier	Winkel and Hudde (1997)
			Clutch size		Increasing	
			Fledging success		Increasing	
Collared flycatcher (*F. albicollis*)	NAO	Yes, +	Laying date	−	Earlier	Both and Visser (2001)
		Yes, +	Laying date	− (between and within individuals)		Przybylo et al. (2000)
			Clutch size	+ (within individuals)		
			Fledging success	0		

Species	Climatic variable		Breeding variable			Reference
Nuthatch (*Sitta europea*)	Temp.		Hatching date	—	Earlier	Winkel and Hudde (1997)
Great tit (*Parus major*)	Temp.	Yes, +	Hatching date	—	Earlier	Winkel and Hudde, (1997)
			Clutch size		0	
			Fledging success		0	
		Variable	Laying date	Variable	0, (earlier in only 5 of 13 pop.)	Visser *et al.* (2003)
Blue tit (*P. caeruleus*)	Temp.	Yes, +	Hatching date	—	Earlier	Winkel and Hudde, (1997)
			Clutch size		0	
			Fledging success		0	
		Variable	Laying date	Variable	0, (earlier in only 3 of 11 pop.)	Visser *et al.* (2003)
Tree swallow (*Tachycineta bicolor*)	Temp.	Yes, +	Laying date	—	Earlier	Dunn and Winkler (1999)
			Clutch size	0	0	Winkler *et al.* (2002)
Barn swallow (*Hirundo rustica*)	NAO	Yes, +	Laying date	0	Curvilinear	Møller (2002)
			Clutch size	+ (first clutches) 0 (second clutches)		
			Breeding success[d]	+ (first clutches) 0 (second clutches)		
Black-throated blue warbler (*Dendroica caerulescens*)	SOI[e]		Young fledged	+		Sillet *et al.* (2000)

Only studies that included both environmental and breeding data are included. + indicates a positive relationship between the climatic variable and the measure of breeding performance; − indicates a negative relationship; 0 indicates no relationship. Blanks indicate no results were reported. Temp. indicates spring temperature.

[a] Temperatures in early and late April have become warmer over time, but mid-April temperatures have tended to decline. Overall, the rate of increase in warming during April has declined.

[b] Sea surface temperature (SST) near the study site during the nestling period.

[c] Large positive values of the North Atlantic Oscillation Index (NAO) are associated with warm, moist winters in coastal Northern Europe, whereas low, negative values are associated with cold, dry winters. The NAO has increased over the past 20–30 years and this has contributed to warmer winters (Hurrell 1995). However, temperatures were not reported at the study site.

[d] Number of fledglings (estimated at 12 d of age) divided by clutch size.

[e] The Southern Oscillation Index (SOI) has large positive values in La Niña years and low, negative values in El Niño years. In this study, food availability during the breeding season was correlated positively with the SOI ($r = 0.58$).

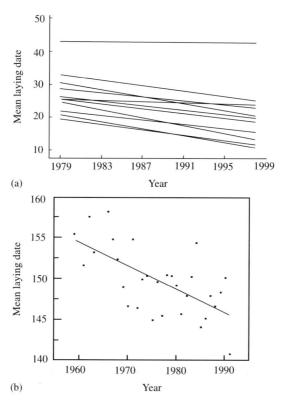

(a)

(b)

Figure 1 Large-scale studies of change in laying dates of (a) blue tits (Visser *et al.*, 2003) and (b) tree swallows (Dunn and Winkler, 1999). Lines are regressions for separate study areas in (a) and North America in (b). Laying dates are expressed in days from 1 April in (a) and from 1 January in (b). Figures are reproduced with permission of the Royal Society of London.

in clutch size have been reported in pied flycatchers (Järvinen, 1989; Winkel and Hudde, 1997) and first broods of barn swallows (*Hirundo rustica*; Møller, 2002), but not in collared flycatchers (Sheldon *et al.*, 2003), great or blue tits (Winkel and Hudde, 1997) or tree swallows (Winkler *et al.*, 2002; see Table 2).

It is not clear why some populations of birds are laying earlier, but clutch size has not increased. In tree swallows, for example, there is a strong negative relationship between clutch size and laying date, and although laying date has advanced by an average of nine days, there has been no significant change in clutch size (Winkler *et al.*, 2002). One possibility is that birds are optimising the combined date of laying, number of eggs laid and days of incubation, so young hatch at the peak of food abundance for nestlings. For example, if climate change is shifting the peak of food abundance earlier, it may be advantageous for females to advance the date of laying in order to maintain synchronisation between the date of hatch and the peak of food for nestlings, but not to make any changes in clutch size (or the length of incubation) that may disrupt the synchrony. Over the past 39 years, the mean laying date of great tits in Oxford, England has advanced in response to climate change, but their mean clutch size has not changed (Cresswell and McCleery, 2003). Nevertheless, these birds have synchronised their hatch dates with the peak date of food abundance by adjusting the length of incubation to annual variations in temperature. Another possible explanation is that climate change has made it easier for some individuals to start laying earlier, but not caused any change in how many eggs they can lay. Winkler *et al.* (2002) found that variation in laying dates was reduced in years with warmer temperatures, and they suggested that warmer temperatures were primarily causing later-laying individuals to advance their laying dates, while early-laying individuals were constrained from laying earlier. There is abundant evidence from field studies that birds laying earlier in the breeding season may risk death or increased energetic costs from periods of cold weather (e.g., Zajac, 1995; Newton, 1998). In most bird species there is much more variation in laying date than clutch size, and food supplementation experiments often result in larger changes in laying date than clutch size, consistent with the idea that there are stronger constraints on clutch size variation.

Climate change could also influence reproductive success through its effects on developing nestlings or fledglings. Warmer temperatures could lead to a greater food supply when parents are feeding nestlings and, hence, greater fledging success. Conversely, it could also produce a mismatch between the timing of breeding and food supply and, consequently, lower fledging success (Visser *et al.*, 1998; Both and Visser, 2001; Møller and Merilä, 2004, this volume). To date, relatively few studies have examined the effects of climate change on fledging success, and the results appear to be variable. Some studies have shown no long-term changes in fledging success (Järvinen, 1989), whereas others have shown long-term declines (Moss *et al.*, 2001) and increases (Winkel and Hudde, 1997; Møller, 2001; see Table 2). The few studies that have extended their analysis to recruitment and population growth also show mixed results. For example, population sizes of white-tailed ptarmigan (*Lagopus leucurus*; Wang *et al.*, 2002), capercaillie (*Tetrao urogallus*; Moss *et al.*, 2001) and tufted puffins (*Fratercula cirrhata*; Gjerdrum *et al.*, 2003) all appear to be declining as temperatures increase. In contrast, changes in climate (warming temperature or increasing NAO or Southern Oscillation Index (SOI)) are associated with higher recruitment or numbers of breeders in golden plovers (*Pluvialis apricaria*; Forchhammer *et al.*, 1998), dippers (*Cinclus cinclus*; Sæther *et al.*, 2000), barn swallows (*H. rustica*; Møller, 2002) and black-throated blue warblers (*Dendroica caerulescens*; Sillett *et al.*, 2000). This variation in the effects of climate on reproductive success suggests that there are ecological and life history differences among species that affect their response to warming temperatures (see below). There may also be density dependent effects on reproductive success that limit changes in population size (Møller *et al.*, 2004, this volume). For example, the population growth of dippers in Norway is influenced by the direct effect of winter temperature on survival, as well as density-dependent effects on recruitment (Sæther *et al.*, 2000). Thus, warmer spring temperatures may lead to earlier breeding and greater production of young, but that higher productivity may be reduced by lower survival or recruitment (see also Wilson and Arcese, 2003).

VI. THE EFFECT OF CLIMATE CHANGE ON PARTICULAR SPECIES AND LOCATIONS

The effects of climate change on timing of breeding have been examined in at least 45 species, most of which were studied in the United Kingdom by Crick and Sparks (1999) (Table 2). Of these 45 species, 27 (60%) showed a significant advancement in laying or hatching date over time. The advancement in the United Kingdom averaged 3 days earlier for every 1°C warmer (range: 1.1–5.5 days/°C; Crick and Sparks, 1999). Thus, the historical record indicates that most species have advanced breeding, and of the 27 that showed an overall advancement, 85% (24/27) also bred earlier during warmer springs, which suggests that climate change has played a role. On the other hand, many species (40%) have not shown any change in laying date. This lack of response could be due to: (1) insufficient climate change at a particular study site (i.e., local differences in temperature change), (2) insufficient climate change for a particular species, in which case further increases in temperature may eventually elicit a response, (3) an inherent lack of response to temperature in some species (e.g., time of breeding may be more closely related to photoperiod or precipitation in some species, especially those in the tropics), or (4) insufficient statistical power to detect a response of the current magnitude.

Although past changes in climate may have been insufficient to elicit a response in some species or regions, future warming is likely to produce stronger changes in phenology. The global average temperature is projected to increase 1.4–5.8°C by 2100 (Houghton *et al.*, 2001), which is significantly more than the 0.6°C increase during the past century. Indeed, Crick and Sparks (1999) predicted that 14 species that previously showed no historical change in laying date will advance their laying by an average of 6.7 days over the next 80 years (based on supplemental material in Crick and Sparks, 1999). Ecological and life history differences might also explain some of the variation in responses among species.

Some species may not show a response to changes in temperature because their timing of breeding is more closely related to photoperiod or rainfall (e.g., Wikelski *et al.*, 2000), which is likely to be especially important in the tropics, or their food supply is influenced less by seasonal changes in air temperature (e.g., fish-eating birds such as herons; Butler, 1993). Furthermore, species with generalist diets may show less of a response to climate change than those with more specialized diets, if the ability to feed on a greater variety of prey items increases the probability that suitable sources of food will be available during breeding. Crick and Sparks (1999) found no obvious associations between the response of laying date to climate change and diet, but their sample of 36 species in the United Kingdom included just five non-passerines, none of which were raptors, whose food supply (fish, small mammals and birds) may be less dependent on temperature.

Some of the difference between species in the response to climate change is related to body size. Using the data of Crick and Sparks (1999), Stevenson and Bryant (2000) found that the long-term change in laying date was smallest in large species and greatest in small species. If temperature constrains early laying, then species with smaller body mass may be more affected by temperature, because they have greater energy expenditure per gram than larger birds (Walsberg, 1983), which increases their relative thermoregulatory costs and may make them more sensitive to temperature (Stevenson and Bryant, 2000). Alternatively, large and small birds may differ in other ways that affect their response. For example, larger species may time their breeding based more on endogenous resources stored before breeding (capital-breeders; Drent and Daan, 1980), whereas smaller species may time their breeding based on rates of resource intake closer to the time of breeding (income breeders). As a consequence, changes in temperature near the time of breeding may be more likely to affect income breeders, which may also be smaller in body mass. To my knowledge, this alternative has not been examined in any detail.

Lastly, differences between species may be related to variation in life history. A recent study of laying dates in 24 populations of blue and great tits throughout Europe found that the response of species was related to regional differences in temperature change, as well as the proportion of second broods in the population. Across six countries, laying dates advanced significantly in just five of 13 great tit populations and three of 11 blue tit populations (Visser *et al.*, 2003). Different factors appeared to explain the responses in different regions. In Russia and Finland, there was no change in laying date apparently because temperatures have not increased. In Corsica, there was no change in laying date because blue tits feed on caterpillars that emerge relatively late in the season and these insects were not affected by changing temperatures earlier in the season. However, even within regions with similar patterns of temperature change (Netherlands and Belgium), there were populations that responded differently. To explain these situations Visser *et al.* (2003) suggested that the response to climate change was linked to the frequency of second broods. They found that in both great and blue tits, laying date advanced more in populations with a stronger decrease in the percentage of second broods over time. They hypothesise that as spring temperatures increase, the caterpillars eaten by tits develop faster, and in populations with mostly single broods, the timing of laying will follow any shifts toward earlier dates of peak food abundance. As the first brood tracks the earlier peak of food abundance, the first clutch will become more valuable to reproductive success, whereas the second brood will become relatively less valuable, because they are less synchronised with the peak of food abundance (Visser *et al.*, 2003). Over time, populations that are tracking changes in food abundance and advancing the laying date of first clutches should also show the largest declines in the proportion of second clutches. This is exactly what Visser *et al.* (2003) found, although they caution that the results are correlational and should be followed by more detailed studies of how this pattern arises. For example, it would be useful to know if, over time, there has been stronger selection against second clutches in populations that have experienced warming temperatures.

Evidence is mounting that the response of species also varies geographically and over time. For example, great tits are laying eggs earlier in Oxford, England (McCleery and Perrins, 1998), but not in the Netherlands (Visser *et al.*, 1998). The English study also found that most of the increase in laying date has occurred since the late 1970s; there was no change in laying date prior to the 1970s. Geographic variation in the response of species has also been reported in tree swallows (Dunn and Winkler, 1999; Hussell, 2003) and pied flycatchers (Sanz, 2003). Climate change had a stronger effect on laying date at western locations within the range of tree swallows and a stronger effect at northern locations within the range of pied flycatchers. Thus, the effects of climate change on breeding phenology may vary depending on the location and time period studied.

These studies show that the same species can differ in its response to temperature across relatively small regions (i.e., regional heterogeneity of response). Other studies show that ecologically similar species may differ in their response to climate even in the same location (i.e., species heterogeneity). For example, reproductive success of sage (*Amphispiza belli*) and Brewer's (*Spizella breweri*) sparrows at the same shrubsteppe site in Oregon, USA, responded differently to precipitation at long- (7 months prior to breeding) and short-term (during breeding) scales (Rotenberry and Wiens, 1991). Reproductive success was greater in both of these species in years with more precipitation during the 7 months prior to breeding (long-term scale). However, the species differed in their reproductive response to weather during the breeding season (short-term). Reproductive parameters of Brewer's sparrows were relatively unaffected by temperature or precipitation during the breeding season, whereas clutch size of sage sparrows was related positively to the level of precipitation during the breeding season and fledging success was related negatively to maximum temperature. Thus, ecologically similar species can respond differently to the same environmental conditions, possibly because they are affected at different temporal scales (short- or long-term).

In summary, the response to climate change may vary between species as a consequence of differences in: (1) local temperature change, (2) diet (generalist or specialist, and how closely the food supply is affected by temperature),

(3) body size which affects thermoregulatory costs, (4) life history (migratory or resident, and the reproductive value of first and second broods; Visser *et al.*, 2003) and (5) the time scale over which species acquire resources for breeding (e.g., income and capital breeders; Drent and Daan, 1980). Another layer of complexity is added when one considers how different species interact with their competitors and parasites. For example, Martin (2001) found that changes in precipitation lead to shifts in microhabitat use by orange-crowned (*Vermivora celata*) and Virginia's (*V. virginiae*) warblers primarily because orange-crowned warblers dominate Virginia's warbler and limit their use of some microhabitats. Studies of additional species are needed to test these hypotheses.

VII. WHAT CAUSES CHANGES IN TIMING OF BREEDING?

To date, studies of the effects of climate change on timing of breeding have necessarily been based on correlations with temperature or other climate variables (NAO, SOI, precipitation). As a consequence, inferences have to be made cautiously. In particular, it is possible that long-term changes in phenology are actually caused by changes in human land use, pollution or changes in breeding density. For example, in a declining population, breeding date may shift earlier if the decline in density leads to relatively greater availability of breeding resources per individual (Dunn and Winkler, 1999). Some of the other possible explanations can be dismissed by careful data analysis. Sanz *et al.* (2003), for example, examined the effects of breeding density on timing of laying and breeding performance of pied flycatchers and found that it had no effect. Similarly, they concluded that any changes in habitat quality were unlikely to produce the particular pattern of breeding performance they observed: laying date and clutch size have not changed over time, but fledging success, fledgling and adult body mass and daily energy expenditure of parents feeding young have all declined. They argue that a change in habitat quality would likely affect clutch size as well as the other measures of breeding performance. Furthermore, they found a long-term increase at their study areas in both temperature (in mid-May) and the normalised difference vegetation index (NDVI), which is related to the level of photosynthetic activity and the timing of oak leafing, which, in turn, is related to the abundance of caterpillars eaten by pied flycatchers (Sanz *et al.*, 2003). Thus, there is a plausible chain of correlations leading back from breeding performance to temperature and climate.

Assuming that timing of breeding is responding to climate change, the next question is what mechanism produces the changes in timing of breeding? Long-term advances in the timing of breeding could be due to: (1) phenotypic plasticity, (2) movements of southern birds, which may be adapted to breed earlier, farther north (i.e., gene flow) or (3) directional selection on laying date of resident birds (Przybylo et al., 2000). Most studies have used cross-sectional data, so it has not been possible to distinguish the responses of individuals under different temperature regimes (phenotypic plasticity) from microevolutionary change (which requires directional selection and additive genetic variation for breeding date). In a 16-year study of collared flycatchers, Przybylo et al. (2000) found evidence that timing of laying was related negatively to the NAO index in analyses of both cross-sectional (between individual) and longitudinal (within individual) analyses. The absence of any long-term change in laying date and the similar results with both types of analyses suggests that individuals are making flexible responses to changing conditions on the breeding grounds (phenotypic plasticity) and there has been no long-term evolutionary change. Whether birds have the ability to respond to climate change is an important issue as it may influence the ability of populations to persist over the next century (Sæther et al., 2004, this volume).

VIII. CONCLUSIONS

Some of the strongest evidence for the effects of climate change on organisms comes from studies of phenology (see reviews by Walthers et al., 2002; Parmesan and Yohe, 2003; Root et al., 2003). In a survey of 677 species of plants and animals, 62% showed trends toward earlier phenology consistent with a warming climate (Parmesan and Yohe, 2003). Studies of birds have made major contributions to understanding the response of animals to climate change, and they will be important in the future for monitoring and understanding the mechanistic basis for phenological change. Although there is a large body of knowledge about the breeding biology of birds, one of the biggest challenges in the future will be to predict how climate change will affect the reproductive performance of different species throughout their ranges. This review has shown that there are many species responding to climate change, but the magnitude of response differs among species and locations in ways that we do not yet understand. In the future, some of the most interesting and useful data will come from large-scale collaborative studies, such as those of tit populations across Europe (Visser et al., 2003). Similar networks of researchers studying other model species (e.g., pied flycatchers, swallows) would be useful for understanding the ecological and life history differences that lead to variation in response to climate change. Although observations by amateurs will continue to be valuable for monitoring climate change (Whitfield, 2001), more coordinated studies of food availability and breeding ecology are now needed on a continental scale, and volunteers are less likely to contribute detailed ecological information. For example, understanding the mechanistic basis of phenological change will require banding large numbers of individuals for longitudinal studies over many years. Europeans have led the way in the study of climate change and the establishment of networks of researchers, but to address climate change at a global scale, we need to expand research networks to other continents, particularly to areas that are warming, but have not yet received much study (e.g., Japan, parts of southern Africa, Australia and South America).

One of the more urgent questions is: what is the relationship between changes in timing of breeding and population dynamics? Many studies have identified cases where species are changing their timing of breeding, but we often do not know what effect (if any) this has on population size. This information will be needed to understand the long-term consequences of climate change. For example, Wilson and Arcese (2003) showed that song sparrows (*Melospiza melodia*) bred earlier on Mandarte Island, British Columbia, in warmer years and earlier breeding led to more fledglings, but greater reproductive success was not associated with subsequent growth of the population, primarily because juvenile recruitment was lower when density was higher. On the other hand, Møller (2002) found that warmer and wetter springs in Denmark (higher NAO index) were associated with larger clutches and more immunocompetent young in the first brood, which subsequently led to more recruits from first broods in the population the next year. More detailed studies such as these are needed on model species in which networks of researchers can collaborate to produce analyses that have large sample sizes and cover broad geographic regions.

ACKNOWLEDGEMENTS

I thank A.P. Møller, J.J. Sanz, L.A. Whittingham and an anonymous reviewer for helpful suggestions on the chapter.

REFERENCES

Arcese, P., Sogge, M.K., Marr, A.B. and Patten, M.A. (2002) (Ed. by A. Poole and F. Gill), *The Birds of North America*, Vol. 704. The Birds of North America, Philadelphia, PA.

Barbraud, C. and Weimerskirch, H. (2001) *Nature* **411**, 183–186.

Bellot, M.D., Dervieux, A. and Isenmann, P. (1991) *J. Ornithol.* **132**, 297–302.

Bijlsma, R.G. (1998) *Limosa* **71**, 137–148.

Both, C. and Visser, M.E. (2001) *Nature* **411**, 296–298.

Brown, C.R. and Brown, M.B. (1999) *Condor* **101**, 230–245.

Brown, J.L., Li, S.H. and Bhagabati, N. (1999) *Proc. Natl. Acad. Sci. USA* **96**, 5565–5569.

Bryant, D.M. (1975) *Ibis* **117**, 180–216.

Butler, R.W. (1993) *Auk* **110**, 693–701.

Cresswell, W. and McCleery, R. (2003) *J. Anim. Ecol.* **72**, 356–366.

Crick, H.Q.P. and Sparks, T.H. (1999) *Nature* **399**, 423.

Crick, H.Q.P., Gibbons, D.W. and Magrath, R.D. (1993) *J. Anim. Ecol.* **62**, 263–273.

Crick, H.Q.P., Dudley, C., Glue, D.E. and Thomson, D.L. (1997) *Nature* **388**, 526.

Daan, S., Dijkstra, C., Drent, R. and Meijer, T. (1988) *Proc. Int. Ornithol. Congr.* **19**, 392–407.

Dijkstra, C. and Zijlstra, M. (1997) *Ardea* **85**, 37–50.

Drent, R.H. and Daan, S. (1980) *Ardea* **68**, 225–252.

Dunn, P.O. and Winkler, D.W. (1999) *Proc. R. Soc. Lond. B* **266**, 2487–2490.

Eeva, T., Veistola, S. and Lehikoinen, E. (2000) *Can. J. Zool.* **78**, 67–78.

Forchhammer, M.C., Post, E. and Stenseth, N.C. (1998) *Nature* **391**, 29–30.

Gjerdrum, C., Vallee, A.M.J., St Clair, C.C., Bertram, D.F., Ryder, J.L. and Blackburn, G.S. (2003) *Proc. Natl. Acad. Sci. USA* **100**, 9377–9382.

Hahn, T.P., Boswell, T., Wingfield, J.C. and Ball, G.F. (1997) *Current Ornithology*, Vol. 14. pp. 39–80, Plenum Press, New York.

Hogstad, O. and Stenberg, I. (1997) *J. Ornithol.* **138**, 25–38.

Hoodless, A.N. and Coulson, J.C. (1998) *Bird Study* **45**, 195–204.

Hotta, M. (1996) *Jpn J. Ornithol.* **45**, 23–30.

Houghton, J.T., Ding, Y., Griggs, D.J., Noguer, M., Vander Linden, P.J., Dal, X., Maskell, K. and Johnson, C.A. (Eds.) (2001) *Climate Change 2001: The Scientific Basis.* Cambridge University Press, Cambridge.

Hurrell, J.W. (1995) *Science* **269**, 676–679.

Hussell, D.J.T. (2003) *Auk* **120**, 607–618.

Järvinen, A. (1989) *Ornis Fenn.* **66**, 24–31.

Jones, L.R. (1986) *Condor* **88**, 91–93.

Lack, D. (1968) *Ecological Adaptations for Breeding in Birds.* Methuen, London.

Martin, T.E. (2001) *Ecology* **82**, 175–188.

McCleery, R.H. and Perrins, C.M. (1998) *Nature* **391**, 30–31.

Meijer, T., Nienaber, U., Langer, U. and Trillmich, F. (1999) *Condor* **101**, 124–132.

Møller, A.P. (2002) *J. Anim. Ecol.* **71**, 201–210.

Møller, A.P. and Merilä, F. (2004) (Ed. by A. P. Møller, W. Fiedler and P. Berthold), *Birds and Climate Change, Advances Ecol. Res.* **35**, 109–128.

Moss, R., Oswald, J. and Baines, D. (2001) *J. Anim. Ecol.* **70**, 47–61.

Myneni, R.B., Keeling, C.D., Tucker, C.J., Asrar, G. and Nemani, R.R. (1997) *Nature* **386**, 698–702.

Nager, R.G. and van Noordwijk, A.J. (1995) *Am. Nat.* **146**, 454–474.

Newton, I. (1998) *Population Limitation in Birds.* Academic Press, San Diego, CA.

Parmesan, C. and Yohe, G. (2003) *Nature* **421**, 37–40.

Perdeck, A.C. and Cave, A.J. (1989) *Ardea* **77**, 99–106.

Perrins, C.M. (1970) *Ibis* **112**, 242–253.

Price, T. and Liou, L. (1989) *Am. Nat.* **134**, 950–959.

Przybylo, R., Sheldon, B.C. and Merila, J. (2000) *J. Anim. Ecol.* **69**, 395–403.

Pulido, F. and Berthold, P. (2004) In: *Birds and Climate Change, Advances Ecological Research* (Ed. by A. P. Møller, W. Fiedler and P. Berthold), Vol. **35**, 149–181.

Rodriguez, C. and Bustamante, J. (2003) *J. Anim. Ecol.* **72**, 793–810.

Root, T.R., Price, J.T., Hall, K.R., Schneider, S.H., Rosenzweig, C. and Pounds, J.A. (2003) *Nature* **421**, 57–60.

Rotenberry, J. and Wiens, J.A. (1991) *Ecology* **72**, 1325–1335.

Saether, B.E., Engen, S., Møller, A.P., Matthysen, E., Adriaensen, F., Fiedler, W., Leivits, A., Lambrechts, M.M., Visser, M.E., Anker-Nilssen, T., Both, C., Dhondt, A.A., McCleery, R.H., McMeeking, J., Potti, J., Røstad, O.W. and Thomson, D. (2003) *Proc. R. Soc. Lond. B* **270**, 2397–2404.

Sæther, B.E., Tufto, J., Engen, S., Jerstad, K., Røstad, O.W. and Skåtan, J.E. (2000) *Science* **287**, 854–856.

Sæther, B.E., Sutherland, W.F. and Engen, S. (2004) (Ed. by A. P. Møller, W. Fiedler and P. Berthold) *Birds and Climate Change, Advances Ecol. Res.* **35**, 183–207.

Sanz, J.J. (2002) *Global Change Biol.* **8**, 409–422.

Sanz, J.J. (2003) *Ecography* **26**, 45–50.

Sanz, J.J., Potti, J., Moreno, J., Merino, S. and Frias, O. (2003) *Global Change Biol.* **9**, 461–472.

Sheldon, B.C., Kruuk, L.E. and Merila, J. (2003) *Evolution* **57**, 406–420.

Sillett, T.S., Holmes, R.T. and Sherry, T.W. (2000) *Science* **288**, 2040–2042.

Stevenson, I.R. and Bryant, D.M. (2000) *Nature* **406**, 366–367.

Thompson, P.M. and Ollason, J.C. (2001) *Nature* **413**, 417–420.

Visser, M.E. and Holleman, L.J.M. (2001) *Proc. R. Soc. Lond. B* **268**, 289–294.

Visser, M.E., van Noordwijk, A.J., Tinbergen, J.M. and Lessells, C.M. (1998) *Proc. R. Soc. Lon. B* **265**, 1867–1870.

Visser, M.E., Adriaensen, F., van Balen, J.H., Blondel, J., Dhondt, A.A., van Dongen, S., du Feu, C., Ivankina, E.V., Kerimov, A.B., de Laet, J., Matthysen, E., McCleery, R., Orell, M. and Thomson, D.L. (2003) *Proc. R. Soc. Lond. B* **270**, 367–372.

Walsberg, G.E. (1983) *Avian Biology*, Vol. 7. Academic Press, New York, pp. 161–220.

Walthers, G.-R., Post, E., Convey, P., Menzel, A., Parmesan, C., Beebee, T.J.C., Fromentin, J.-M., Høgh-Guldberg, O. and Bairlein, F. (2002) *Science* **416**, 389–395.

Wang, G., Hobbs, N.T., Giesen, K.M., Galbraith, H., Ojima, D.S. and Braun, C.E. (2002) *Climate Res.* **23**, 81–87.

Whitfield, J. (2001) *Nature* **414**, 578–579.

Wikelski, M., Hau, M. and Wingfield, J.C. (2000) *Ecology* **81**, 2458–2472.

Wilson, S. and Arcese, P. (2003) *Proc. Natl. Acad. Sci. USA* **100**, 11139–11142.

Winkel, W. and Hudde, H. (1997) *J. Avian Biol.* **28**, 187–190.

Winkler, D.W., Dunn, P.O. and McCulloch, C.E. (2002) *Proc. Natl. Acad. Sci. USA* **99**, 13595–13599.

Winter, S.W., Gorlov, P.I. and Andryushchenko, Y.A. (1999) *Vogelwelt* **120**, 367–376.

Zajac, T. (1995) *Acta Ornithol.* **30**, 145–151.

Global Climate Change Leads to Mistimed Avian Reproduction

MARCEL E. VISSER,* CHRISTIAAN BOTH AND
MARCEL M. LAMBRECHTS

I. SUMMARY

Climate change is apparent as an advancement of spring phenology. However, there is no *a priori* reason to expect that all components of food chains will shift their phenology at the same rate. This differential shift will lead to mistimed reproduction in many species, including seasonally breeding birds. We argue that climate change induced mistiming in avian reproduction occurs because there is a substantial period between the moment of decision making on when to reproduce and the moment at which selection operates on this decision. Climate change is therefore likely to differentially alter the environment of decision-making and the environment of selection. We discuss the potential consequences of such mistiming, and identify a number of ways in which either individual birds or bird populations potentially can adapt to reproductive mistiming.

E-mail address: M.Visser@nioo.knaw.nl (M. Visser)

ADVANCES IN ECOLOGICAL RESEARCH, VOL. 35
0065-2504/04 $35.00 DOI 10.1016/S0065-2504(04)35005-1

II. INTRODUCTION

Different aspects of global climate change, such as the increase in ambient temperature during the last 30 years, have been shown to influence a wide range of biological systems (Wuethrich, 2001; Walthers *et al.*, 2002). One important aspect of biological systems that has been affected by climate change is phenology, such as the timing of reproduction. For many avian species in the temperate zone, there is only a short period in the annual cycle when conditions are most suitable for reproduction. An increase in ambient temperature most likely leads to an advancement of optimal breeding conditions, and as a consequence birds are expected to advance their timing of reproduction.

The impact of climate change on timing of reproduction has frequently been reported in correlational studies which show that laying dates have advanced in the last decades in many bird species (Crick *et al.*, 1997; Crick and Sparks, 1999; Parmesan and Yohe, 2003; Root *et al.*, 2003; Dunn, 2004, this volume). However, recent investigations revealed considerable variation in responses of breeding time to climate change both within and among avian species (Dunn, 2004, this volume). Parmesan and Yohe (2003) reported that 78 out of 168 species of birds advanced their laying date (47%) but 14 (8%) showed a delay and the other 76 (45%) showed no significant change. Dunn and Winkler (1999) showed that tree swallows (*Tachycinta bicolor*) differ in the advancement of egg laying date across North America and Visser *et al.* (2003a) showed in a European wide comparative study that some great (*Parus major*) and blue tit (*P. caeruleus*) populations advanced their average onset of egg laying during the last 20 years, but others not (see also Figure 1a; Dunn, 2004, this volume, and Sanz (2002) for geographical variation in how the North Atlantic Oscillation (NAO) affects great and blue tit laying dates). Also laying date in *Ficedula* flycatchers differed across Europe, and this variation correlates very well with variation in changes in spring temperature; populations occurring in areas without warming of spring did not advance their laying date, while the more the local temperature increased the more the birds advanced their laying date over the years (Both *et al.*, 2004).

Clearly, there is variation both within and among species in how much, if at all, the timing of reproduction has advanced. The most relevant question is however not *whether or not* a population has advanced the timing of reproduction *per se*, but rather whether bird populations have shifted their timing of egg laying *sufficiently* to match the shift in the period of favourable conditions for raising chicks. This has been only rarely considered when reporting changes in timing of reproduction in the context of climate change. The few cases where advancement of both favourable breeding conditions and avian reproduction has been investigated show insufficient shifts in the timing of reproduction (great tit, Visser *et al.*, 1998, pied flycatcher, Both and Visser, 2001). These studies reported a mismatch between the time of maximum availability of food for raising chicks and the time the chicks are fed by the parents (Figure 1). Within populations there have always been individuals that were mistimed, such as late arriving immigrant great tits which lay too late (Nager and van Noordwijk, 1995), or blue tits in evergreen forests which lay too early as a consequence of gene-flow from deciduous to evergreen habitat (Dias *et al.*, 1996; Blondel *et al.*, 2001). However, climate change weakens the

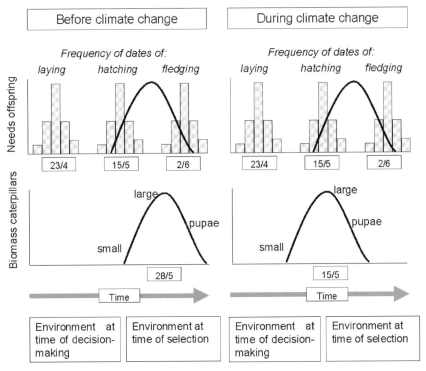

Figure 1   A graphical representation of reproductive mistiming due to climate change in Dutch great tits: left panels prior and right panels during climate change. Top panels represent the frequencies of (from left to right) laying dates, hatching dates and fledging dates. The need for food for the chicks in the nest is indicated with a solid line. Laying dates have not changed under climate change. Lower panels represent the biomass of defoliating caterpillar (main prey for the nestlings) availability: initially low as there are many, but very small, caterpillars, then a peak at the time when there are large caterpillars, followed by a decline when caterpillars start to pupate and are no longer available as prey. The peak date in caterpillar biomass shifts to an earlier date due to climate change, and there is no longer synchronisation between the time the nestlings are fed and maximum food abundance: the population is mistimed. Below the lower panels the environments of decision-making and selection are indicated (Figure 2).

synchronisation between food availability and offsprings' needs for the average individual in the population.

In this chapter we will explain why we expect that climate change in general will lead to reproductive mistiming in birds (Section III). Next, we will discuss the consequences of this mistiming for population numbers (Section IV) and how birds may adapt to mistiming, either via responses at the individual or population level (Section V). Throughout the chapter we will illustrate our arguments with our own research on blue tits, great tits and pied flycatchers, as reproductive mistiming in the context of climate change has most extensively been studied in these bird species.

III. WHY GLOBAL CLIMATE CHANGE WILL LEAD TO REPRODUCTIVE MISTIMING

Birds are adapted to year-to-year variation in the timing of favourable conditions, i.e., in general they lay earlier in warmer springs (Dunn, 2004, this volume). However, often birds cannot use direct measurements of abundance of the food fed to nestlings to time their reproduction, as gonad development and laying eggs occurs well before the date when chicks hatch. Therefore, birds need to use cues to time their laying date, i.e., environmental variables at the time of egg formation (the environment of decision-making). These cues should have a predictive value for when food is plentiful later in the season (the environment of selection that determines the contribution to the following generation, c.f. van Noordwijk and Muller, 1994). Different cues may be used, that are combined and weighted to produce a physiological response mechanism translating the cues from the environment into a laying date (Lambrechts and Visser, 1999). As the environment differs from year-to-year, and consequently the value of the cues differs, birds also lay at different times.

A serious but often ignored aspect of global climate change is that temperatures (or other weather variables) have not just simply increased, but that temperatures in some periods change at a different rate than in other periods, or that temperatures at different locations (wintering versus breeding area) are changing in a different way (Visser et al., 1998, 2003b; Inouye et al., 2000; Walthers et al., 2002). This means that the cues (the environment of decision-making) are affected in a different way by climate change than the environmental variables that affect the timing of favourable conditions (the environment of selection), and that climate change will lead to mistiming, i.e., that the change in timing of the birds is unequal to the change in timing of their main food sources for chick feeding (Figure 2).

Next we will discuss how climate change will alter the environment of decision-making (Section A) and the environment of selection (Section B), and

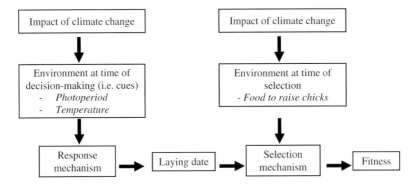

Figure 2   A schematic outline on how climate change may affect reproduction in birds. When the environment of decision-making is affected this may, via the response mechanism, affect laying date while changes in the environment of selection will affect the fitness consequences of laying dates.

therefore how it will influence the time that birds lay eggs and/or the conditions during which chicks are raised (Figure 2). We argue that there is no *a priori* reason why these two environments should change at the same rate in response to climate change, and why this likely leads to reproductive mistiming in birds (Section C).

A. Changes in the Environment of Decision-Making

Birds have to use those cues in the environment of decision-making that have some predictive value for the environment of selection in order to lay their eggs at the appropriate time (Wingfield *et al.*, 1992; van Noordwijk *et al.*, 1995). Climate change may well affect the environment of decision-making but assessing the magnitude of this effect is hampered by our deficient knowledge of the birds' response mechanism; which cues are used and how are these cues integrated to initiate gonadal growth and egg production. Only after this is known can we determine to what extent these cues are altered by climate change.

An important cue for the timing of reproduction is photoperiod as this sets the window within which reproduction will take place (Silverin *et al.*, 1993; Wingfield, 1993; Gwinner, 1996; Lambrechts and Perret, 2000; Sæther *et al.*, 2004, this volume). But as the change in photoperiod is the same every year, this cue cannot play a role in how birds adapt to year-to-year variation in optimal breeding time and thus additional cues should play a role. The fact that climate change will not affect such an important cue as day length may already limit the response in laying date to changes in climate. The strength of this effect depends on how climate change influences the supplementary cues species use to time reproduction. Understanding adaptation to climate change thus should focus on these supplementary cues, although it is important to know how the effect of day length is constraining a response to climate change.

Ambient temperature is one of the most likely candidates for a supplementary cue. Many temperate zone species, including Great and Blue Tits, lay earlier in warmer springs (Kluyver, 1951; van Balen, 1973; Dhondt and Eyckerman, 1979; Perrins and McCleery, 1989; Dunn, 2004, this volume). However, to what extent gonad development and/or egg laying are directly sensitive to ambient temperature is unclear (Silverin, 1995, see for a review Visser and Lambrechts, 1999 and Dunn, 2004, this volume). The bud burst development of trees, which could serve as an accurate measure for the timing of the food peak does not play a direct role in the timing of gonad development and egg laying, as experimentally shown for great and blue Tits (Visser et al., 2002).

Opportunistic breeders that breed after the start of a temporal unpredictable food supply do not have a window of reproduction set by photoperiod. They use cues that directly predict when food will be abundant, independent of calendar date. Zebra finches (*Taeniopygia guttata*) use flushes of grass seed as a cue, which themselves are directly related to the unpredictable incidence of effective rainfall (Zann, 1999). Also red crossbills (*Loxia curvirostra*, Hahn, 1998) and Piñon jays (*Gymnorhinus cyanoephalus*, Ligon, 1974) use the availability of their food supply directly to decide when to start egg laying. For these granivorous species there is a strong correlation between the environments of decision-making and selection, because of the highly predictable timing of their food supply at the time of egg laying. Therefore, they might be well able to cope with sudden changes in the timing of food availability due to climate change.

Long distance migrants have an extra handicap to adjust their breeding date to climate change, because on the wintering grounds it is often impossible to predict changes in the onset of optimal reproductive conditions on the breeding grounds. They use internal clocks or cues such as day length to time the start of spring migration (Gwinner, 1996), and this constrains their adjustment to climate change (Both and Visser, 2001; Coppack and Pulido, 2004, this volume).

Up to now we discussed the cues in the environment of decision-making used by birds, and how these cues may facilitate or complicate the development of appropriate proximate responses to climate change. But reproducing females also have to gather large amounts of proteins to produce eggs and in early spring food resources are often limited. Often these resources for egg production are different from those used to raise nestlings. We need to know what resources females use during egg production and whether the phenology of these resources is affected by climate change. For instance, Dutch great tits provide their nestlings mainly with caterpillars from oaks

(*Quercus robur*) while, as far as we know, they use insects from birch (*Betula pubescens*) and larch (*Larix deciduas*) in the egg formation period. Oak bud burst is strongly temperature sensitive while birch and larch are not, and thus the interval between budburst of these species and that of the oak has become shorter over the past two decades (Visser *et al.*, 1998).

In conclusion, whether changes in the environment of decision-making will lead to changes in the timing of reproduction will vary among species. Opportunistic breeders are likely to be affected by climate change during the period of gonad development and egg formation, while species that migrate are less likely to be affected at that time. For resident species, the question whether changes in the environment of decision-making will lead to changes in the timing of reproduction will strongly depend on the importance of day length relative to other cues. As day length is not changing it is likely that in species in which this cue has an overriding effect, timing of egg laying will be affected to a minor extent.

B. Changes in the Environment of Selection

How a bird species should adjust its breeding date in response to climate change depends to a large extent on the response of other parts of the food chain during the time of selection. In a strongly seasonal environment that is affected by temperature dependent processes, we expect that the phenology of a large part of the ecosystem should advance in response to climate change. In temperate regions, invertebrates such as caterpillars consuming leaves of deciduous trees provide a good example (Buse *et al.*, 1999). Both tree and insect phenology are temperature dependent and only young leaves are palatable to most herbivorous insects eaten by birds. In general birds specialising on these insects produce only a single successful brood per season. To ensure successful reproduction, these single-brooded bird species should adjust their timing of raising chicks (breeding date) to advances in insect availability. However, once the birds have started egg laying, they cannot lay more than one egg a day, and are rather fixed in the duration of incubation and chick rearing (van Noordwijk *et al.*, 1995). Therefore, an increase in ambient temperature starting after the onset of egg laying will not lead to an advancement of the chick stage. By contrast, the development of tree leaves and their herbivorous invertebrates is strongly temperature dependent, and an increase in temperature after the start of egg laying in birds thus advances the peak in food availability without the birds being able to respond.

In environments with a less pronounced peak in food availability during the stage of chick feeding, and where species can raise more than one brood per season, the need to adjust breeding date to climate change may be less strong, although also in these environments individuals with multiple broods will have the highest fitness and therefore should time their reproduction appropriately. Furthermore, a seasonal decline in offspring value is observed in many species (Nilsson, 1999 for a review) forcing birds to breed as early as possible. In several raptor species prey (rodent) populations steadily increase during the summer, but the seasonal decline in offspring survival selects for birds to breed as early as possible (Daan et al., 1988, 1990). The need to advance breeding date in response to climate change is probably less in species relying on food sources for which the availability is not temperature dependent, such as granivorous species. Again, this depends on other date dependent effects on fitness of both parents and offspring, and competition for available food. In general we need to know the effect of climate change on development and availability of essential resources such as nestling food. Therefore, we need to develop a more multi-trophic approach including all underlying parts of the food chain to make a good prediction which species are most likely to be affected by climate change.

Ecological differences, even on a small spatial scale, can affect variation in the environment of selection and consequently how birds should be reacting to climate change. Blue tits on the island of Corsica breed in two very different types of habitats that are interspersed on short distance; evergreen and deciduous Oaks (Lambrechts et al., 1997a,b; Blondel et al., 1999). The phenology of the evergreen habitat (Pirio population) is a late bud burst and a late food peak. In contrast, the leafing of trees in the deciduous habitat (Muro population) and the food peak is much earlier. The blue tit populations in these forests are well adapted to these two species of oaks, with the Pirio population laying about 1 month later than the Muro population, even though these habitats are only 25 km apart. Given this long interval, it may well be that climate change will affect these two populations very differently. As suggested by Visser et al. (2003a) the populations in habitats with a late food peak, which also breed late, may escape from impacts of climate change either because they breed outside the seasonal window during which climate change occurs, or because the phenology of the entire food chain is less temperature sensitive. The latter explanation might be less likely as it has been suggested that there is no difference in temperature sensitivity in gonadal development between early and late breeding blue tit populations in the Mediterranean region (Lambrechts et al., 1997, 1999).

In conclusion, the environment of selection is to a large part determined by food availability when chicks are in the nest. Climate change is expected to change the environment of selection for many species, but perhaps most strongly for species depending on a relatively short peak in resources. As the environment of selection strongly depends on the underlying levels of the food chain, including the vegetation, there may be very fine-scale spatial variance in changes in this environment.

C. Changes in Synchronisation

Great tits start their egg laying about 1 month before they need most food during the nestling stage, and the environments of decision-making and selection are thus separated in time. Both environments are not necessarily exposed to the same changes in climate, and such a differential change leads to a disruption in synchronisation between the nestling time and the time food availability is at its maximum. There is no *a priori* reason to expect that the time of egg formation in the environment of decision-making and the time and speed of insect growth in the environment of selection shift at the same rate in response to climate change. The phenology of the nestlings' food is determined by the underlying levels of the food chain, for instance by the vegetation. These organisms differ strongly from birds and are likely to have different response mechanisms to time their phenology and growth patterns. Especially the importance of the non-changing photoperiod cue is very different as it plays an important role in the response mechanism of many temperate zone avian species but hardly affects the timing of maximum insect availability. Although natural selection has favoured avian response mechanisms to lead to a similar advancement and delay in warm and cold spring as the phenology of the underlying tropic levels, it is important to realise that this only holds for the environment in which these response mechanisms have evolved. Thus, birds will react to cues that have predictive value of when the food peak will be, but these cues probably only work in a restricted climatic range and rely on a certain temporal structure in these cues. Problems arise when climate change shifts temperatures outside the normal range, and when some periods in the season are more affected than others (Visser *et al.*, 2003b).

Climate change may lead to differential changes in the breeding dates and the time of maximum food abundance. The breeding date may become later than the best time to rear the chicks, as discussed earlier, but birds may also advance more than the food peak. This phenomenon has been reported in the British great tit population of Wytham Wood, Oxford, England, where the birds advanced the egg laying date stronger than the peak time of their caterpillar prey (Cresswell and McCleery, 2003). This makes an interesting contrast with the Dutch great tit population on the Hoge Veluwe where there has been a shift in the peak date of caterpillar availability but not in the average onset of egg laying (Visser *et al.*, 1998, 2003b). As a consequence, there has been an increased selection for early laying in the Hoge Veluwe population, but not in Wytham Wood where selection on early laying has declined. Thus, the Dutch great tits responded too weakly to climate change while in the UK population the birds, which in the past always bred late relative to the food peak, now on average got better synchronised with their prey. This better synchronisation at Wytham Wood also gives the birds more flexibility to advance hatching in warm springs, and consequently the breeding success has increased over the years. As outlined above, there is no *a priori* reason to expect that synchrony is maintained. Synchronisation may get better or worse, either because the phenology of the birds advances less or more than the phenology of the food they need. Why the one situation occurs in the UK population and the other in the Dutch population remains unclear. The explanation that in the UK temperatures in both the pre-laying and the breeding period has increased (Stevenson and Bryant, 2000) while in the Netherlands only the breeding period has become warmer (Visser *et al.*, 1998) was rejected in a comparison of laying date trends on a Europe wide scale (Visser *et al.*, 2003a) as within Europe there are areas where temperatures in the pre-laying period have not increased but laying date has advanced (two populations in Belgium and one in the Netherlands).

Another example of how climate change leads to mistiming comes from a long-distance migrant, the pied flycatcher. Long distance migrants may have even more difficulties in maintaining synchronisation with their food sources. They normally arrive at their breeding grounds only shortly before they start breeding, which constrains their ability to anticipate the advancement of their food sources. This was clearly observed in a Dutch pied flycatcher population that advanced its egg laying date by about 7 days reducing the time interval between arrival and laying (Both and Visser, 2001). Since the food peak in the same area advanced by about 2 weeks, the synchronisation between the nestling period and the food peak deteriorated, and as a consequence selection for early breeding increased (Both and Visser, 2001). The Dutch flycatchers thus advanced their laying date, but the rate of advancement was insufficient to track change in the environment of selection, and this is because their arrival date on the breeding grounds has not advanced, constraining their advancement in laying date.

IV. CONSEQUENCES OF REPRODUCTIVE MISTIMING

Mistiming will have consequences, both for the life-histories and population dynamics of birds. A clear example of the consequences of disrupted synchrony between realised and optimal breeding time is that food availability is lower for parents when feeding their young (Sanz *et al.*, 2003). Thomas *et al.* (2001a) have clearly demonstrated the negative consequences of such mistiming for parents. They measured metabolic effort of blue tit parents facing variation in prey availability during the chick stage (Figure 3). Especially in poor evergreen oak habitat in continental southern France, Blue tit parents face mismatching between nestling demand and prey availability, forcing them to increase foraging effort beyond their sustainable limit (Drent and Daan, 1980), thereby potentially influencing adult survival in these habitats. These negative consequences were not observed in a Corsican blue tit population that is nicely adapted to the same

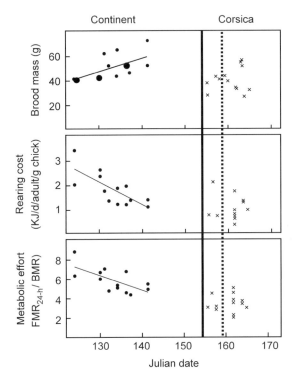

Figure 3   Consequences of mistiming in French blue tits: top panel gives the fledging mass of the offspring, middle panel the rearing cost for the parents and bottom panel the metabolic effort for the parents, all against the date at which the chicks are 8 days old (the age at which their need for food peaks). Symbols on the left hand of the graphs (•) are for a continental population, symbols on the right (x) are for a Corsican population. The grey lines represent the date at which the abundance of food used to rear the offspring peaks (solid line for the mainland, broken line for Corsica) (adapted from Thomas *et al.*, 2001a,b).

habitat type (Thomas *et al.*, 2001a), and are probably also less pronounced in very rich habitats with super abundant food (Thomas *et al.*, 2001b; Tremblay *et al.*, 2003).

   The population consequences of mistiming in the Dutch flycatchers are clearly visible as a decline in population numbers in some, but not all habitats. During the last decades flycatchers breeding in nest boxes declined dramatically in rich deciduous forests, while no systematic trend was observed in mixed and pure pine forests (Both *et al.*, in preparation; Figure 4). This is because the food peak has become earlier and shorter in those rich deciduous forests, and flycatchers used to breed in this habitat during the declining phase of the food peak. The birds

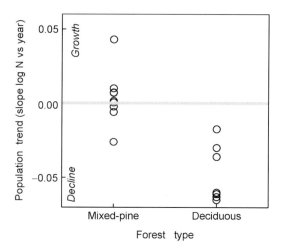

Figure 4 Changes in population numbers (slope of log population number versus year for 1987–2001) for 16 Dutch pied flycatcher populations in either mixed-pine or deciduous forests (adapted from Both *et al.*, in preparation).

however did not track the advancement of this food peak and as a consequence they now miss this food peak completely, and their numbers have declined. In mixed and pine forests the food peak is later and probably broader, and although the food peak has advanced here as well, the birds were better able to adjust their laying dates to this advancement. However, as we showed for the mixed forest of the Hoge Veluwe, the fitness landscape changed here as well, because the birds were unable to fully adjust their laying date to the advancement of the food peak. Since further adjustment to climate change in these flycatcher populations is constrained by their arrival time, we expect that birds also would miss the food peak in this habitat if the advancement continues, with similar population declines as now observed in deciduous habitats.

In the Dutch great and blue tit studies we do not yet see a decline in population numbers (unpublished results). This might be because winter food conditions have an overriding effect on population numbers, mainly via the survival probability for birds in their first winter (Perdeck *et al.*, 2000; Visser et al., submitted). One important food source in winter is beech mast, which occurs on average once every three winters, and drives the population dynamics of the tits to a large extent (Perrins, 1966). Interestingly, the amount of beech crop is predicted to increase over the next century as a consequence of climate change (van der Meer *et al.*, 2002), which might counteract the negative effect of disrupted synchrony. However, if the mistiming continues to increase then also for the tit species, there will be a moment when also the earliest birds hatch their chicks too late to make use of the high but short peak in caterpillar biomass abundance, similar to what we have seen in pied flycatchers.

V. ADAPTATIONS TO REPRODUCTIVE MISTIMING

Given the negative consequences of reproductive mistiming it is important to assess the way individuals or populations can adapt to climate change. Here we will address whether birds can adjust other components of their life history to reduce the negative fitness consequences of mistiming (Section A), and whether populations will adapt to the changing environment and thus restore the match between the timing of maximal food requirements and the time of maximal food availability (Section B).

A. Responses of Individuals

In many bird species, individuals reproduce earlier in warm than in cold springs (Dunn, 2004, this volume). However, as we have argued above, climate change might lead to an insufficient advancement of laying date in these warm springs and the birds might be mistimed. After birds have laid their first egg, they have only limited possibilities for adjustment of hatching dates, which determines the mismatch between offspring needs and food abundance (van Noordwijk et al., 1995; Wesolowski, 2000; Visser et al., 2003b). One option is laying a smaller clutch to advance the hatching date with one or a few days.

However, often the opposite is found: when birds advance their laying date, they also lay a larger clutch (Winkel and Hudde, 1997; Winkler et al., 2002; Both and Visser, submitted). This could be explained by inflexibility of the generally found decline of clutch size with laying date. However, as Both and Visser (submitted) show, this decline is not fixed but is for pied flycatchers steeper in warmer springs, because in those years a clear fitness cost of being late exists. Birds thus trade clutch size against hatching date in order to maximise their fitness (see also Dunn, 2004, this volume).

An alternative to advance hatching date for a given laying date is to reduce the interval between the last egg being laid and the start of incubation. Great tits (Visser et al., 1998, 2003b) and pied flycatchers (Both and Visser, submitted) at the Hoge Veluwe (NL) have over the past two decades started to incubate incomplete clutches, thereby reducing the incubation period. It is likely that such an early start of incubation will come at a cost, as it will lead to hatching asynchrony, and perhaps to an increased mortality risk for the last hatched chicks. Interestingly, the great tits at Wytham Wood (UK) got on average better synchronised with the environment over the years and their incubation time has increased, resulting in reduced hatching asynchrony and higher nest survival (Cresswell and McCleery, 2003). Rather than just using laying date to adjust hatching date, birds may use the whole complex of laying date, clutch size and start of incubation to advance the chick stage in response to climate change.

Birds may learn from previous experience and thereby adapt to disrupted synchrony with the environment. Great tits that hatched their chicks late relative to the peak in food abundance advanced their timing the next year (Nager and van Noordwijk, 1995). Learning the best breeding date is causally linked to experienced mistiming as shown experimentally in blue tits by Grieco et al. (2002). This learning mechanism may have evolved to cope with spatial and temporal variation in food phenology and birds may become better adapted during their life if the climate changes directionally. Although beneficial to the individual bird, in short lived species we predict that this learning effect will be insufficient to prevent the population from adverse effects of mistiming, because most birds are young and have no prior experience and hence lay too late.

In migratory species, the timing of arrival at the breeding grounds may hamper advancement of laying date (Both and Visser, 2001). Migrants may be able to advance their arrival date as the improved circumstances en route may speed these birds up to a certain extent, but this would only give a limited flexibility to climate change if the start of migration is very rigid (Coppack and Both, 2002). Most species of long distance migrants arrive at the breeding grounds some time before they start breeding in an average year, which gives them some flexibility to anticipate natural variation in the advancement of spring between years (Drent et al., 2003). By reducing the time between arrival and breeding they can to a limited extent adjust to the advancement of the optimal breeding time caused by climate change. However, if arrival is not advanced, their adjustment to climate change is insufficient in the long term (Both and Visser, 2001). A change in migration schedule is therefore needed for adapting to a changing climate. This probably requires selection on the response mechanism for the onset of migration, and that the environmental conditions at the wintering grounds and en route allow earlier migration. Since climate change differs between latitudes this may constrain selection for earlier migration.

B. Population Responses

If food sources advance more than the birds' timing of breeding, early laying birds will increasingly produce more surviving offspring relative to late laying birds. This will lead to negative selection differentials for laying date (i.e., selection for earlier laying). The question is whether such increased selection will lead to changes at a population level such that mistiming will be reduced. To answer this question we first need to establish where selection will operate. Selection for the timing of breeding operates not on laying dates as a genetically "fixed" trait, but on how an individual (or genotype) adjusts its laying date to environmental cues. This reaction norm or physiological response mechanism gives an individual the flexibility to adjust its laying date to the prevailing circumstances so that it lays at the approximate optimal date for a whole range of years differing in environmental conditions. It is important to realise that natural selection has shaped this mechanism for a specific set of abiotic variables (weather patterns). Global climate change is not only an increase in average annual temperatures, but temperatures in some periods increase more than in other periods of the year. As a consequence, climate change may disrupt the particular set of correlated environmental variables under which the physiological response mechanisms have evolved. This would be no problem if the lower levels of the food chain have the same physiological response mechanisms and are thus affected similarly as birds. However, they are likely to have other response mechanisms and are thus affected by different environmental variables. These different variables in turn might be affected differently by climate change (Visser and Holleman, 2001). It is therefore likely that under climate change the physiological response mechanisms underlying phenotypic plasticity in laying dates no longer accurately predict the timing of favourable conditions, and are thus no longer adaptive.

As the response mechanisms underlying phenotypic plasticity are no longer adaptive, the only way birds can adapt to these changes in climate is via selection on the shape of the reaction norm, i.e., on the genetic basis of phenotypic plasticity. This makes the debate on whether the observed advancement of laying date is due to phenotypic plasticity or a shift in gene frequencies of limited value (Przybylo *et al.*, 2000; Both and Visser, 2001). As reaction norms evolved for a limited range of environmental conditions, it is not the question whether animals can adapt either via phenotypic plasticity or via changes in gene frequencies, but much more via changes in the genetic basis of phenotypic plasticity (Visser and Holleman, 2001).

Synchrony could be restored by a change in gene frequencies of the response mechanism of laying dates to environmental circumstances (phenotypic plasticity). Such an evolutionary response is, however, only possible if egg laying date is heritable. There is some evidence for that (Merilä and Sheldon, 2001), but the conventional idea about heritability may not be very useful for a plastic trait like laying date that individuals alter depending on the environmental conditions that they encounter. Natural selection will operate on the response mechanism underlying this plasticity (see above and Coppack and Pulido, 2004, this volume). We only have a good idea how this mechanism is shaped for a few species, but we have no idea of variation in this mechanism even in these species, let alone whether this variation has a heritable component. To complicate matters further, selection on plasticity of a character (on the reaction norm) is believed to be slow (van Tienderen and Koelewijn, 1994). This makes it unlikely that natural selection will be able to keep up with rapid changes in climate.

The response to selection also depends on the scale of climate change and on the extent of gene flow across local populations. Even on a small scale there is variation in the rate at which populations shift their laying date (Visser *et al.*, 2003a). There will be gene flow between some of these populations. An extreme example of contrasting selection pressures on a very local scale, over which dispersal can easily occur, is the blue tit on Corsica. Two populations that are only 25 km apart evolved pronounced differences in laying date in response to local optimal breeding time, which is assumed to have a genetic basis. This suggests that evolution in egg laying dates is possible even at a small spatial scale (Lambrechts *et al.*, 1997a,b; 1999; Blondel *et al.*, 1999), especially in conditions where selection pressures are predictable at the scale (e.g., landscape) of the dispersal range of the species.

Gene flow could, however, not only hamper adaptation but also allow new genes with different response mechanisms to enter the population and thereby broaden the range of phenotypes on which selection can act. We have limited evidence that there is spatial variation in the response mechanism, both on large (Silverin *et al.*, 1993; see Figure 2 in Coppack and Pulido, 2004, this volume) and small geographical scales (Lambrechts *et al.*, 1997a). While there may have been strong selection against gene flow in the past, as this counteracts local adaptation in the current situation of disrupted synchrony, selection may actually favour gene flow. Immigration of birds with different response mechanisms, some of which are now better adapted to the local situation than the residents, may allow populations to adapt to climate change (Coppack and Both, 2002).

VI. DISCUSSION AND CONCLUSIONS

Climate change is already apparent as an advancement of spring phenology. We argued in this chapter, however, that there is no *a priori* reason to expect that all components of food chains will shift their phenology at the same rate. The main reason is that the different components in a food chain will have different response mechanisms underlying the timing of their phenology and that apart from an overall change in temperature, there will also be a change in weather patterns (correlations between climatic variables, either in time or space). We may expect that under undisturbed weather patterns different response mechanisms of components in the food chain will be selected so that they shift more or less to the same degree with varying spring conditions. However, this expectation will no longer hold under novel weather patterns, simply because the mechanisms have not been selected under these new environments.

When different components of the food chain shift at different rates, this will lead to mistiming and we believe that such mistiming resulting from climate change will be a general phenomenon. If we now zoom in from this general picture to avian reproduction (Figure 2), birds seem to be especially vulnerable. The environment at the time they produce their eggs (environment of decision-making) is in general much earlier than the environment when selection will occur on for instance synchrony between offspring needs and prey availability (the environment of selection). The evolved response mechanisms are appropriate for the range of prevailing conditions, and climate change is a trend that will at first fall within the normal range of temperatures. In the short term, an increase in temperatures therefore may allow birds to cope with their existing reaction norms. If these temperatures fall outside the normal range, or if periods in spring differ in their temperature change, the prevailing reaction norms become maladaptive. Furthermore, the photoperiod (which is not affected by climate change) is an important component of the environment of decision-making for birds, but not of the environment of selection (see also Coppack and Pulido, 2004, this volume). As a result, birds are unlikely to advance their laying date at an appropriate rate.

There are only three studies that compared the actual shift in laying date with the one that would be optimal. Two of these (great tit, Visser et al., 1998; pied flycatcher, Both and Visser, 2001) find that the shift in laying date is insufficient, despite the fact that there is a significant advancement of laying date in the pied flycatcher. The third (great tit, Cresswell and McCleery, 2003) finds that selection in the past favoured early laying birds, but that this early bird advantage declined with time, which may suggest that birds advanced more than their food source.

We have assumed almost throughout the entire chapter that birds respond to climate change by changes in their laying date solely. But for the pied flycatcher and we also investigated clutch size and onset of incubation (Both and Visser, submitted). We concluded that the pied flycatchers may, rather than just using laying date as a way to advance hatching date, use the whole complex of laying date, clutch size and start of incubation. Birds may also adjust other correlated life-history traits as some species show no change in laying date but do respond in whether or not they make a second brood (i.e., produce within a season a second brood after a successful first brood has fledged). In Northwest European great tit populations a dichotomy existed, with populations that did not advance their laying date reducing the proportion of second broods over time, while populations with a stable and low frequency of second broods advancing their laying date (Visser et al., 2003a).

Birds may actually not just time when to start their first brood in a season but may be optimising the timing of their entire annual cycle. This is particularly likely as many life-history traits affect each other, like for instance reproduction and moult or autumn migration and moult (Coppack et al., 2001).

For long-distance migrants this inter-dependence of timing of life-history traits is even more clear as their spring arrival date, and thus their spring departure date and migration speed, strongly limits their advancement of laying date and influences fitness consequences of mistiming (Coppack and Both, 2002). Thus viewing the impact of climate change on reproduction in isolation from other life-history events may be convenient but perhaps misleading. Species may respond to climate change, but in different life-history characters than timing of reproduction. Thus, when we compare species in their shift in laying date we might find that some of the non-advancing populations change a different life-history character. On the other hand, some of the advancing populations may not be doing so well at all if it turns out that they are advancing but not at the appropriate rate. This makes the interpretation of the meta-analysis on shifts in timing of avian reproduction difficult (Crick *et al.*, 1997; Crick and Sparks, 1999; Parmesan and Yohe, 2003; Root *et al.*, 2003).

Traditionally, birds were thought to time their breeding season so that the peak demand of the chicks matches the peak availability of prey used to feed the chicks (Lack, 1933). Although this synchronisation is indeed one of the major selection pressures acting on timing of reproduction, more recent studies have emphasised that also egg laying and incubation are costly, and therefore act as important selection pressures when a bird should lay (Monaghan and Nager, 1997; Stevenson and Bryant, 2000; Visser and Lessells, 2001). And as costs made in an early part of the reproductive cycle also affect costs in subsequent parts (Heany *et al.*, 1995), selection might operate on the entire reproductive cycle, or even on the entire life cycle when also timing of moult or migration is included, making it harder to identify the environment of selection.

Several important questions need to be addressed in future research on causes and consequences of climate change. A first outstanding question to be examined is whether our prediction that climate change will in general lead to mistiming holds. For this, more species should be investigated. Visser *et al.* (2003b) provide a number of methods to evaluate this: change in selection pressures on egg laying dates, direct measurements of changes in phenology of resources for offspring, and examination of behavioural responses such as an earlier in onset of incubation.

A second outstanding question is whether natural selection will lead to changes at the population level so that mistiming will be reduced. A prerequisite for this is that there is heritable variation in the response mechanisms birds use. We know very little about this, but a first step would be to document existing variation in response mechanisms within species having a wide geographic distribution (using any of the 75 species with a distribution from Scandinavia to Southern Europe), as Silverin (1995) (see Figure 2 in Coppack and Pulido, 2004, this volume) did for the great tit. If there is sufficient geographical variation in response mechanisms, this indicates that species can adapt to a wide range of climatic conditions that is at least as large as the predicted changes in climate. This argument assumes that climate change will not lead to climatic conditions which fall clearly outside the existing range in climates encountered throughout the distributional range of species. We would predict that populations with a wide geographical distribution are better able to cope with climate change as they will have ample intraspecific variation in response mechanisms, so that dispersal across latitudes may allow rapid adaptation through changes in existing frequencies of response mechanisms. If selection would act on response mechanisms we would predict genetic change at the level of the response mechanism, and therefore need to estimate heritable variation in response mechanisms. As endocrinologists and quantitative geneticists have detailed knowledge of the processes underlying timing of reproduction, we advocate close collaboration between endocrinologists, geneticists and ecologists to unravel these response mechanisms, and its genetic basis.

A third outstanding question is the consequences of mistiming for population dynamics. We know very little about this at present except for recent work on the pied flycatcher in the Netherlands, where the populations are disappearing in rich oak forests (Both *et al.*, in preparation; see Section IV). Dunn (2004, this volume) discusses the effect of climate change on population processes such as fledging success but to what extent mistiming underlies the effects he reports is unknown. These questions are just starting to be addressed (Sæther *et al.*, 2004, this volume).

A final objective for future research is to increase our understanding of the effect of climate change on phenology for a number of species that differ in ecology. Currently, this is greatly hampered by our limited knowledge of on the one hand the causal mechanisms underlying laying date and on the other hand by our knowledge of exactly what the underlying selection pressures are. Knowledge of the causal mechanism (the response mechanism) is crucial, as this will indicate which changing environmental variables are relevant. It will also enable us to make predictions about how timing will change, if there are no genetic changes in the population, over the next century (using the International Panel for Climate Change predictions). Knowledge of selection pressures, including those on the birds' prey (thus using a multi-trophic approach), is also crucial as this enables us to assess to what extent laying dates should shift. Only when we have such a yardstick can we answer the question whether populations have advanced their laying date *sufficiently* to match the shift in the period of favourable conditions for reproduction rather than just *whether or not* a population has advanced. We predict, and perhaps fear, that in most cases we will find that the shift in laying date has not been sufficient and hence that climate change will have led to mistiming.

ACKNOWLEDGEMENTS

We thank Barbara Helm and Anders Pape Møller for their comments on a previous version of this chapter. MEV and CB thank J.H. van Balen who kept the long-term study on the Hoge Veluwe going for many years and J. Visser for managing the databases. We thank the board of the National Park "de Hoge Veluwe" for their permission to work within their reserve. MML thanks the Montpellier group for their continuous effort to gather data in Mediterranean environments.

REFERENCES

Blondel, J., Dias, P.C., Perret, P., Maistre, M. and Lambrechts, M.M. (1999) *Science* **285**, 1399–1402.
Blondel, J., Perret, P., Dias, P.C. and Lambrechts, M.M. (2001) *Genet. Sel. Evol.* **33**, 121–139.
Both, C. and Visser, M.E. (2001) *Nature* **411**, 296–298.
Both, C., Artemyev, A.V., Blaauw, B., Cowie, R.J., Dekhurzen, A.J., Eeva, T., Enemar, A., Gustafsson, L., Ivankina, E.V., Järvinen, A., Metcalfe, N.B., Nyholm, N.E.I., Potti, J., Ravussin, P.-A., Sanz, J.J., Silverin, B., Slater, F.M., Sokolov, L.V., Török, J., Winkel, W., Wright, J., Zang, H. and Visser, M.E. (2004) *Proc. R. Soc. Lond.* **271**, 1657–1662.
Buse, A., Dury, S.J., Woodburn, R.J.W., Perrins, C.M. and Good, J.E.G. (1999) *Funct. Ecol.* **13(suppl.)**, 74–82.
Coppack, T. and Both, C. (2002) *Ardea* **90**, 369–378.
Coppack, T. and Pulido, F. (2004) (Ed. by A.P. Møller, W. Fiedler and P. Berthold), *Birds and Climate Change. Advances Ecol. Res.* **35**, 129–148.
Coppack, T., Pulido, F. and Berthold, P. (2001) *Oecologia* **128**, 181–186.
Cresswell, W. and McCleery, R. (2003) *J. Anim. Ecol.* **72**, 356–366.
Crick, H.Q.P. and Sparks, T.H. (1999) *Nature* **399**, 423–424.
Crick, H.Q.P., Dudley, C., Glue, D.E. and Thomson, D.L. (1997) *Nature* **388**, 526.
Daan, S., Dijkstra, C., Drent, R.H. and Meijer, T. (1988) *Proceedings of the 19th International Ornithol. Congress*, Ottawa 1986, 392–407.
Daan, S., Dijkstra, C. and Tinbergen, J.M. (1990) *Behaviour* **114**, 83–116.
Dhondt, A.A. and Eyckerman, R. (1979) *Ibis* **121**, 329–331.
Dias, P.C., Verheyen, G.R. and Raymond, M. (1996) *J. Evol. Biol.* **9**, 965–978.
Drent, R.H. and Daan, S. (1980) *Ardea* **68**, 225–252.
Drent, R.H., Both, C., Green, M., Madsen, J. and Piersma, T. (2003) *Oikos* **103**, 272–294.
Dunn, P. (2004) (Ed. by A.P. Møller, W. Fedler and P. Berthold), *Birds and Climate Change, Advances Ecol. Res.* **35**, 67–85.
Dunn, P.O. and Winkler, D.W. (1999) *Proc. R. Soc. Lond. B* **266**, 2487–2490.
Grieco, F., van Noordwijk, A.J. and Visser, M.E. (2002) *Science* **296**, 136–138.
Gwinner, E. (1996) *Ibis* **138**, 47–63.
Hahn, T.P. (1998) *Ecology* **79**, 2365–2375.

Heany, V. and Monaghan, P. (1995) *Proc. R. Soc. Lond. B* **261**, 361–365.

Inouye, D.W., Barr, B., Armitage, K.B. and Inouye, B.D. (2000) *Proc. Natl Acad. Sci. USA* **97**, 1630–1633.

Kluyver, H.N. (1951) *Ardea* **39**, 1–135.

Lack, D. (1933) *Proc. Zool. Soc. Lond.* 231–237.

Lambrechts, M.M. and Perret, P. (2000) *Proc. R. Soc. Lond. B* **267**, 585–588.

Lambrechts, M.M. and Visser, M.E. (1999) In: *Proceedings of the 22nd International Ornithol. Congress, Durban* (Ed. by N.J. Adams and R.H. Slotow), pp. 231–233. BirdLife South Africa, Johannesburg.

Lambrechts, M.M., Blondel, J., Hurtrez-Boussès, S., Maistre, M. and Perret, P. (1997a) *Evol. Ecol.* **11**, 599–612.

Lambrechts, M.M., Blondel, J., Maistre, M. and Perret, P. (1997b) *Proc. Natl Acad. Sci. USA* **94**, 5153–5155.

Lambrechts, M.M., Perret, P., Maistre, M. and Blondel, J. (1999) *Proc. R. Soc. Lond. B* **266**, 1311–1315.

Ligon, J.D. (1974) *Nature* **250**, 80–82.

Merilä, J. and Sheldon, B.C. (2001) *Curr. Ornithol.* **16**, 179–255.

Monaghan, P. and Nager, R.G. (1997) *Trends. Ecol. Evol.* **12**, 270–274.

Nager, R.G. and van Noordwijk, A.J. (1995) *Am. Nat.* **146**, 454–474.

Nilsson, J.-A. (1999) In: *Proceedings of the 22nd International Ornithol. Congress, Durban* (Ed. by N.J. Adams and R.H. Slotow), pp. 234–247. BirdLife South Africa, Johannesburg.

Parmesan, C. and Yohe, G. (2003) *Nature* **421**, 37–42.

Perdeck, A.C., Visser, M.E. and van Balen, J.H. (2000) *Ardea* **88**, 99–106.

Perrins, C.M. (1966) *Brit. Birds* **59**, 419–432.

Perrins, C.M. and McCleery, R.H. (1989) *Wilson Bull.* **101**, 236–253.

Przybylo, R., Sheldon, B.C. and Merilä, J. (2000) *J. Anim. Ecol.* **69**, 395–403.

Root, T.L., Price, J.T., Hall, K.R., Schneider, S.H., Rosenzweig, C. and Pounds, J.A. (2003) *Nature* **421**, 57–60.

Sæther, B.E., Sutherland, W.J. and Engen, S. (2004) (Ed. by A.P. Møller, W. Fiedler and P. Berthold), *Birds and Climate Change, Advances Ecol. Res.* **35**, 183–207.

Sanz, J.J. (2002) *Global Change Biol.* **8**, 409–422.

Sanz, J.J., Potti, J., Moreno, J., Merino, S. and Frías, O. (2003) *Global Change Biol.* **9**, 461–472.

Silverin, B. (1995) *Am. Zool.* **35**, 191–202.

Silverin, B., Massa, R. and Stokkan, K.A. (1993) *Gen. Comp. Endocr.* **90**, 14–22.

Stevenson, I.R. and Bryant, D.M. (2000) *Nature* **406**, 366–367.

Thomas, D.W., Blondel, J., Perret, P., Lambrechts, M.M. and Speakman, J.R. (2001a) *Science* **291**, 2598–2600.

Thomas, D.W., Blondel, J., Perret, P., Lambrechts, M.M. and Speakman, J.R. (2001b) *Science* **294**, 471a.

Tremblay, I., Thomas, D., Lambrechts, M.M., Blondel, J. and Perret, P. (2003) *Ecology* **84**, 3033–3043.

van Balen, J.H. (1973) *Ardea* **61**, 1–93.

van der Meer, P.J., Jorritsma, I.T.M. and Kramer, K. (2002) *Forest Ecol. Manag.* **162**, 39–52.

van Noordwijk, A.J. and Muller, C.B. (1994) In: *Animal Societies; Individuals, Interactions and Organisation* (Ed. by P.J. Jarman and A. Rossiter), pp. 180–194. Kyoto University Press, Kyoto.

van Noordwijk, A.J., McCleery, R.H. and Perrins, C.M. (1995) *J. Anim. Ecol.* **64**, 451–458.

van Tienderen, P.H. and Koelewijn, H.P. (1994) *Genet. Res. Camb.* **64**, 115–125.

Visser, M.E. and Holleman, L.J.M. (2001) *Proc. R. Soc. Lond. B.* **268**, 289–294.

Visser, M.E. and Lambrechts, M.M. (1999) In: *Proceedings of the 22nd International Ornithol. Congress, Durban* (Ed. by N.J. Adams and R.H. Slotow), pp. 249–264. BirdLife South Africa, Johannesburg.

Visser, M.E. and Lessells, C.M. (2001) *Proc. R. Soc. Lond. B.* **268**, 1271–1277.

Visser, M.E., van Noordwijk, A.J., Tinbergen, J.M. and Lessells, C.M. (1998) *Proc. R. Soc. Lond. B.* **265**, 1867–1870.

Visser, M.E., Silverin, B., Lambrechts, M.M. and Tinbergen, J.M. (2002) *Avian Sci.* **2**, 77–86.

Visser, M.E., Adriaensen, F., van Balen, J.H., Blondel, J., Dhondt, A.A., van Dongen, S., du Feu, C., Ivankina, E.V., Kerimov, A.B., De Laet, J., Matthysen, E., McCleery, R.H., Orell, M. and Thomson, D.L. (2003a) *Proc. R. Soc. Lond. B* **270**, 367–372.

Visser, M.E., Both, C. and Gienapp, P. (2003b) *Acta Zool. Sinica*, in press.

Walther, G.-R., Post, E., Convey, P., Menzel, A., Parmesan, C., Beebee, T.J.C., Fromentin, J.-M., Høgh-Guldberg, O. and Bairlein, F. (2002) *Nature* **416**, 389–395.

Wesolowski, T. (2000) *J. Orn.* **141**, 309–318.

Wingfield, J.C. (1993) *Gen. Comp. Endocrinol.* **92**, 388–401.

Wingfield, J.C., Hahn, T.P., Levin, R. and Honey, P. (1992) *J. Exp. Zool.* **261**, 214–231.

Winkel, W. and Hudde, H. (1997) *J. Avian. Biol.* **28**, 187–190.

Winkler, D.W., Dunn, P.O. and McCulloch, C.E. (2002) *Proc. Natl Acad. Sci. USA* **99**, 13595–13599.

Wuethrich, B. (2001) *Science* **287**, 793.

Zann, R.A. (1999) In: *Proceedings of the 22nd International Ornithol. Congress, Durban* (Ed. by N.J. Adams and R.H. Slotow), pp. 265–278. BirdLife South Africa, Johannesburg.

Analysis and Interpretation of Long-Term Studies Investigating Responses to Climate Change

ANDERS P. MØLLER* AND JUHA MERILÄ

I. SUMMARY

Effects of recent climate change in mean phenotypic values of different traits in wild bird populations are already apparent, but we are still largely ignorant about the mechanism underlying these changes. Likewise, numerous methodological issues complicating the interpretation of the observed patterns have emerged, but few of these have been widely recognised. Here, we review some of these problems inherent to long-term studies of wild bird populations in the context of understanding climate change effects on mean phenotypes in populations. In particular, we focus on methodological issues such as problems arising from sampling bias, as well as problems of differentiating between phenotypic plasticity, genetic adaptation and gene flow as causes of change in the mean phenotype of a population over time.

E-mail address: amoller@snr.jussieu.fr (A.P. Møller)

ADVANCES IN ECOLOGICAL RESEARCH, VOL. 35
0065-2504/04 $35.00 DOI 10.1016/S0065-2504(04)35006-3

II. INTRODUCTION

Climate of the earth has changed dramatically during the last few decades (e.g., Easterling *et al.*, 1997), and the rate of this change has been among the most extreme in the recent history of life (Lehikoinen *et al.*, 2004, this volume). Given that most phenomena in nature are strongly linked to local climatic conditions, it is not surprising that the signature of this climate change is already apparent in a number of animal and plant populations (Walther *et al.*, 2002; Parmesan and Yohe, 2003; Root *et al.*, 2003).

Many long-term population studies of birds have provided evidence of effects of climatic change on timing of migration, reproduction, population size, and distribution during breeding and winter (Parmesan and Yohe, 2003; Root *et al.*, 2003). However, other studies have found little or no effects (Lehikoinen *et al.*, 2004, this volume; Visser *et al.*, 2004, this volume; Møller *et al.*, 2004, this volume; Böhning-Gaese and Lemoine, 2004, this volume). Therefore, although the general patterns of response are clear, there is a great deal of heterogeneity in these patterns demanding an explanation. While it is possible that the observed heterogeneity may indeed arise for biological reasons (e.g., Visser *et al.*, 2003), it is also possible that methodological differences between studies contribute to such heterogeneity. For instance, apparently different responses in different populations or species might be observed simply because of differences in the methods employed. As a corollary, it is also possible that similar responses are observed although the underlying processes are entirely different. For instance, similar change in mean breeding time of a population in response to climate warming can be attained through phenotypic plasticity, genetic adaptation or migration of new genotypes into a given population. Likewise, although there is no doubt about the scientific value of long-term population studies *per se*, we should not forget that long-term studies also may have some inherent problems that need to be taken into account when drawing inferences.

Here, we discuss potential problems inherent to the interpretation of long-term data, and in particular, how these might serve as an explanation of heterogeneity in documented responses to climate change. These problems range from the choice of variables, to problems of sampling and analysis, and problems of interpretation of potential mechanisms and assumptions underlying the inference drawn. Our aim is not to be critical of particular studies or methods employed, but rather to present a broad overview of the problems that may be encountered in observational long-term studies of wild bird populations. It is only by considering inherent problems in the way in which we conduct research that we can improve scientific enquiry and levels of understanding.

We start out by presenting a number of methodological considerations. These include questions of which phenological and climatic variable to use, the effects of differences in capture probability on estimates of phenotypic variables, and questions about representativeness of the data collected. In the second part, we discuss ways to distinguish between change in phenotype due to phenotypic plasticity and genetic change. In addition, we emphasise that gene flow may cause a temporal change in phenotype by yet another mechanism. Finally, we provide a checklist of crucial questions that may allow better results obtained from long-term data sets.

III. METHODOLOGICAL CONSIDERATIONS

A. Choice of Phenological Variables

Many studies have investigated first arrival date, first breeding date or other measures of earliness in terms of phenology. A smaller number of studies have investigated means, medians or other measures of centrality of frequency distributions. The approach of using first dates has a number of potential problems. First, with only a single observation contributing to the estimate, sampling errors for first dates are considerably larger than for means or medians. Second, measures of first dates will depend much more strongly on changes in overall sampling effort than means or medians. Third, if population size changes during the study period, this will directly affect the probability of encountering an early individual because a smaller number of individuals is sampled in a decreasing population (Tryjanowski and Sparks, 2001).

As an example of this problem, consider the pattern in Figure 1 showing the relationship between mean and first observation for spring arrival in a 32-year time series for the barn swallow *Hirundo rustica* (Møller, 2004 and unpubl.). Although there was a positive relationship between the two measures of spring arrival ($r_s = 0.532, z = 2.964, p = 0.003$), there was also considerable scatter (Figure 1c). In fact, for a given first arrival date mean arrival could vary by as much as 10 days among years. While mean arrival date showed a temporal trend towards earlier arrival ($r_s = -0.376, z = 2.095, p = 0.036$; Figure 1b), that was not the case for first dates ($r_s = 0.099, z = 0.554, p = 0.580$; Figure 1a). However, it should not be forgotten that both first dates and mean dates contain slightly different information. A decrease in mean date, but not in first date could be indicative of a change in the variance, with fewer extreme arrival dates being recorded in recent years. Conversely, a decrease in first arrival date, but no change in mean date could be indicative of

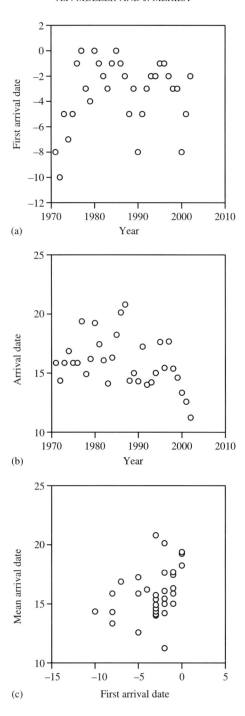

a change in variance of arrival date (see also Lehikoinen *et al.*, 2004, this volume).

There are a number of solutions to this problem of using first dates. First, for a population estimate it might be better to rely on mean values or other measures of centrality. Second, if first dates are investigated, perhaps because no other data were available, it will be crucial to investigate the effects of potential biases, including, e.g., changes in population size or sampling effort with time. Third, estimates of first dates would be less subject to sampling error if based on the mean of a large number of observations, either from a number of amateurs recording dates in a single area, or from several populations within a larger area (e.g., Forchammer *et al.*, 2002). This approach can also be used to assess the accuracy of these variables and to estimate actual among-site variation (and thus representativeness of local results). Such data have been routinely collected and published by local ornithological clubs at least in Fennoscandia, UK and several Central European countries for decades. Cross-validation based on a data set of first dates and mean first dates could provide information on the extent of bias due to sampling when relying on first dates only.

B. Choice of Climate Variables

Most studies in the literature reporting relationships between phenotype and climate are based on simple correlation or time series analyses. Numerous studies have used the local mean temperature for a short period as an index of climatic conditions (e.g., Mason, 1995). Many others have relied on indicators of climatic conditions across large areas such as the North Atlantic Oscillation (NAO), the Arctic Oscillation, or the El Niño Southern Oscillation (ENSO; e.g., Sillett *et al.*, 2000; Aanes *et al.*, 2002; Forchammer *et al.*, 2002). However, there is often little or no explicit justification for the choice of particular variables, raising questions about the appropriateness of choices.

Weather in large areas is determined by predominant patterns of pressure passages as illustrated by the NAO, which is defined as the normalised pressure difference between Stykkisholmur, Iceland and Ponta Delgado, Azores (Hurrell, 1995). Positive values imply frequent passage of low pressures at high latitudes in Scandinavia giving rise to relatively warm winter and spring with heavy

---

Figure 1 Arrival date to the breeding site for yearling male barn swallows from Denmark. (a) Mean arrival date in relation to first arrival date. (b) Arrival date of the first individual during 1971–2002. (c) Mean arrival date during 1971–2002. Each observation represents the value for a single year. Adapted from (Møller 2004 and unpubl.)

precipitation and high pressures with sunny and dry weather in Southern Europe (Hurrell, 1995). Negative values produce the opposite patterns. For migrants, individuals will encounter areas where there are both positive and negative relationships between temperature, precipitation and NAO during migration. Large NAO values will produce low precipitation and high temperatures during spring migration in Southern Europe for many migrants, but heavy precipitation in the breeding grounds in Scandinavia. Therefore, analyses of migratory birds based on NAO may not be easily interpretable since the same climatic conditions may have both positive and negative effects in different areas that even could cancel out.

Migratory birds comprise a very large group of species that live in different places at different times of the year. Thus, environmental conditions during breeding, migration and winter may all on their own—but also in combination—affect the phenotype of migratory birds. Existence of such effects can only be investigated if the relevant climatic conditions for the different parts of the annual cycle are known and available for investigation. This would allow statistical tests to be set up *a priori*. Therefore, appropriate analyses can only be made when there is a thorough knowledge of the migration routes and the wintering grounds of the population under study. For example, a long-term study of a barn swallow population from Denmark revealed a significant change in the mean spring arrival date during the course of 32 years (Figure 1b and c; Møller, 2004 and unpubl.; Møller and Szép, 2004, this volume). The relationship between mean arrival date and NAO was very weak and not statistically significant (Figure 2a), while the relationship with mean temperature at the breeding grounds was only slightly stronger, but still not statistically significant (Figure 2b). However, the best predictor of spring arrival date was environmental conditions in Algeria, North Africa, during spring migration (Figure 2c; $r_s = 0.793, z = 3.269, p = 0.0011$). A similar, but weaker relationship was found for first arrival dates ($r_s = 0.665, z = 2.744, p = 0.0061$). Change in mean arrival date from one generation to the next could be predicted by environmental conditions in Northern Africa (Figure 3). Spring in Northern Africa is the critical period since the birds are emaciated after crossing the Saharan Desert. Survival rate is high during favourable years, allowing many individuals of poor phenotypic quality to continue migration to the breeding areas, thereby leading to a delayed mean spring arrival in Denmark. In adverse years survival rates are low, and only individuals in prime condition survive, advancing the mean spring

---

Figure 2   Mean ($\pm$ S.E.) arrival date to the breeding site by yearling male barn swallows from Denmark in relation to (a) North Atlantic Oscillation index, (b) mean April temperature (°C), and (c) Normalised Difference Vegetation Index (Prince and Justice, 1991) during spring migration in March–May in Algeria. Adapted from Møller, 2004 and unpubl.; Moller and Szép, 2004.

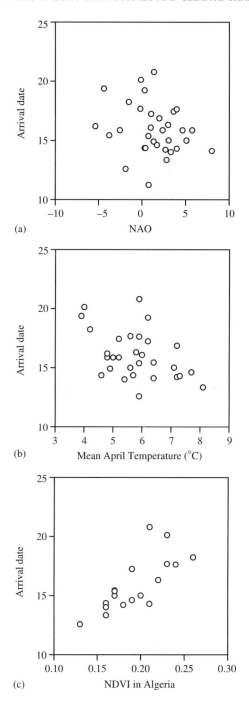

(a) NAO

(b) Mean April Temperature (°C)

(c) NDVI in Algeria

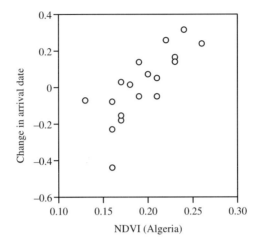

Figure 3   Change in mean arrival date from one year to the next, measured in units of standard deviations, for yearling male barn swallows at Kraghede during the period 1971–2002 in relation to Normalised Difference Vegetation Index (Prince and Justice, 1991) in Algeria during March–May of the previous year. Adapted from Møller (unpubl.).

arrival considerably. Analyses solely based on weather conditions on the breeding grounds would obviously be unable to identify such a carry-over effect from spring migration to the breeding areas. Hence, it is important to note that the lack of correlation between a climatic condition proxy and phenotypic traits of interest should not be interpreted as evidence for lack of an association. It might just as well reflect the situation that a given weather proxy may not be relevant in a given analysis, and that the wrong weather variable has been studied, again emphasising the fact that thorough knowledge of the biology of a species is needed in order to make appropriate statistical tests.

C. Choice of Statistical Models

Apart from the problem of identifying the relevant climatic variable, long-term studies face the problem of identifying the relevant time-window from where the possible climatic effects stem. Time series data are often auto-correlated, i.e., there is a statistical dependence between successive data points. This is also the case in indices of general weather patterns such as the NAO and the El Niño Southern Oscillation (e.g., Hurrell, 1995; Palmer and Pearson, 2003). Thus, conditions in subsequent years are more similar than in years that are separated by longer periods of time. Time series are traditionally analysed using time series analysis (e.g., Box and Jenkins, 1976) to determine the extent to which effects are immediate rather than delayed. Delayed effects may be more likely when organisms have long life-span that will allow accumulation of negative (or positive) effects over many years, or when the delayed effects influence population parameters, e.g., through delayed density-dependence (Beckerman *et al.*, 2002) or evolutionary change. At the individual level, evidence for delayed or long-lasting effects of environmental conditions experienced in the past are overwhelming (reviews in Rossiter, 1996; Lindström, 1999; Metcalfe and Monaghan, 2001). Similar evidence is also available from population level analyses (Beckerman *et al.*, 2002). Consequently, if the time-lags are important, formal time-series analyses could be a useful addition to the toolbox of avian ecologists working on climate change problems.

D. Accounting for Variation in Recapture Probabilities

A number of different mechanisms can explain a change in the mean phenotype of a population over time (see below), but apparent changes in mean phenotype are also possible when no real change has actually taken place. Such changes might be observed for instance if sampling effort during the study period changes, and if individual capture probabilities are associated with individual differences in condition. In turn, such variables are likely to be associated with, for instance, timing of migration (Møller, 1994) and breeding (Price *et al.*, 1988; Sheldon *et al.*, 2003). Therefore, apparent changes in the population mean phenotype might reflect changes in capture probability rather than real changes in the mean phenotype. Capture−mark−recapture analyses (Burnham *et al.*, 1987; Lebreton *et al.*, 1992; White and Burnham, 1999) provide a rigorous framework for estimating demographic and phenotypic variables both before and after selection, while adjusting for differences in capture probability. Surprisingly, few avian studies have relied on these standard techniques to obtain reliable estimates of selection on the variables of interest (but see Larsson *et al.*, 1998; Cooch, 2002; Monros *et al.*, 2002). Biases due to changes in capture probability are less likely to be a problem in studies where the recapture probabilities exceed 90−95%, because almost all birds are included in the analyses, although such studies are very rare. Hence, we suggest that long-term studies investigating changes in mean phenotypic traits over time should seek to verify that the observed changes, or their lack thereof, are not attributable to changes in sampling effort in the course of the study.

E. Representativeness of the Data

A potentially important concern relates to the question as to what degree the data collected from a particular population is representative for the situation in other populations of the same species. Scientists typically work in prime quality habitats to increase the feasibility of their study, while the dynamics of populations are determined by factors, which may differ among habitats. The relative importance of among-habitat dynamics cannot be assumed to remain constant because differences in the distribution and abundance of habitats may affect the probability of emigration and immigration. Depending on whether a prime or a poor habitat is studied, this may give rise to completely different interpretations of temporal change in phenotype as influenced, for instance, by dispersal. A potential solution to this problem is to estimate dispersal and recruitment rates using mark–recapture analyses (Lebreton *et al.*, 1992) and subsequently investigate these estimates for temporal trends. If such trends are identified, these variables could be used as covariates in other analyses to control for temporal trends in the dynamics of populations. We are unaware of such studies.

A related problem may arise because dispersal varies non-randomly across habitats. A higher degree of dispersal from poor than high quality habitats may create the impression that survival rate is decreasing (or that the negative effects of density-dependence are dwindling), while in actual fact survival has remained constant. Direct or indirect effects of climate on the distribution and abundance of habitats may even further complicate this issue. Consequently, an important consideration in all population studies is to ask how representative is the study for the region and/or species at large. In this perspective, combining information from multiple populations as done now in many studies (e.g., Sanz, 2002, 2003; Visser *et al.*, 2003) will provide means for stronger inference than studies based on single populations only. In single population studies, usage of capture–mark–recapture models can help to identify temporal changes in rates of immigration, emigration and recapture probability.

IV. INTERPRETATIONAL CONSIDERATIONS

Three different mechanisms may give rise to directional change in time in the mean phenotype of a population. These include (1) phenotypic plasticity, (2) micro-evolutionary change and (3) gene flow from genetically distinct populations (Hoffmann and Parsons, 1997). All these mechanisms, which are not mutually exclusive, may cause a shift in the mean phenotype of a population and serve as a means of coping with climate change (Hoffmann and Parsons, 1997). Consequently, our understanding of causes and consequences of temporal trends in population mean phenotypes or the absence thereof requires knowledge of the underlying mechanisms. In what follows, we will briefly discuss these three possible explanations for temporal changes in population mean phenotypes, and problems associated with distinguishing between them (see also Sæther et al., 2004, this volume; Böhning-Gaese and Lemöine, 2004, this volume).

A. Phenotypic Plasticity

A change in the mean phenotype of a population can be caused by plastic response to a given environment. Changing environmental conditions over time can create a directional change in the mean phenotype of a population through effects on individual ontogeny or simply, when individuals change their phenotype (e.g., breeding time) in response to changes in the environment. In most published studies, no attempt has been made to differentiate between genetic and purely plastic changes as explanations for the observed changes. Analysing data on variation in individual responses to climatic conditions as indexed by NAO, Przybylo et al. (2000) demonstrated that phenotypic plasticity provides means of responding to climatic variation in collared flycatchers *Ficedula albicollis*. Subsequent analyses of similar data on individual responses to environmental conditions in other species (e.g., Møller, 2002; Schiegg et al., 2002) confirmed that birds possess the ability to respond rapidly to changing environmental condition by means of phenotypic plasticity. We suggest that comparison of individual and population level changes in phenotypic traits as done by Przybylo et al. (2000) should be a useful way to gain insight into the question as to what degree the observed changes in mean phenotypic traits are caused by plasticity or genetic adaptation. An interesting and yet unanswered question in this context is whether there are limits to phenotypic plasticity. Phenotypic plasticity as a means for adjusting to environmental changes may be possible only within a certain range of conditions encountered by a given species. In the face of continuing directional environmental change, limits to plasticity could be met and extant norms of reaction could become maladaptive (evolutionary traps) (see Pulido and Berthold, 2004, this volume).

B. Genetic Change

By creating new selection pressures, climate change can also change the genetic constitution of a population given that there is sufficient heritable variation in the traits that are targets of selection (Lynch and Walsh, 1998). The expected response (R) of a:trait to selection is given by the "breeders equation" (Lynch and Walsh, 1998)

$$R = h^2 S \qquad (1)$$

where h^2 is the heritability of the trait, and S the standardised directional selection differential measuring the strength of selection on the trait.

This equation, or its multivariate extension (see Lynch and Walsh, 1998), will predict the population response to selection given certain assumptions (Lynch and Walsh, 1998; Merilä et al., 2001a). Evolutionary response to selection in avian migratory traits has been documented in several selection experiments (Berthold et al., 1990; Pulido et al., 1996, 2001), but detection of similar responses in the wild are more problematic. One way to distinguish between genetic response and phenotypic plasticity as an explanation for phenotypic change is to show that the observed change over time is predictable from Eq. (1) as demonstrated, e.g., in one species of Darwin's finches (Grant and Grant, 1989). However, even if correspondence between observed and expected responses is found, it may be difficult to exclude phenotypic plasticity as an alternative explanation for the observed change, except when the observed response exceeds that to be expected to be achieved by plasticity alone (see above). Another possibility is to estimate individuals' expected breeding values (Lynch and Walsh, 1998), which are twice the expected deviance in phenotype between trait values for recruits from the mean value in the population. A change in the population's mean breeding value over time would be predicted by Eq. (1). This approach has recently been employed in several studies of wild bird populations (e.g., Kruuk et al., 2001; Merilä et al., 2001a,b; Sheldon et al., 2003) and mammals (e.g., Kruuk et al., 2002b). Finally, common garden experiments in which animals are brought from the wild to standardised environmental conditions can also allow distinction between genetic and environmental effects, since only genetic effects will be maintained across generations in such an experimental set-up (see Coppack and Pulido, 2004, this volume).

However, apart from classic studies of Darwin's finches (Grant, 1986; Grant and Grant, 1989), we are aware of very few examples of a micro-evolutionary response to selection actually occurring as a response to a change in climatic conditions in birds (see also Böhning-Gaese and Lemoine, 2004, this volume). In fact, Merilä et al. (2001a) pointed out that in many long-term studies of vertebrates, heritable traits under apparent directional selection have not changed over time as predicted by quantitative genetic theory. Possible explanations for this discrepancy between theory and data include for instance: (i) biased estimates of heritability and intensity of natural selection, as well as (ii) opposing effects of genetic and environmental factors on trait means (see Merilä et al. (2001a) for full account of possible explanations). Here, we address each of these issues briefly in the context of expected responses to climate change in avian populations.

First, biased estimates of heritability and intensity of natural selection might give us false expectations about the rate and magnitude of phenotypic changes in the wild. As discussed below, heritability estimates are sensitive to environmental conditions under which they are measured, and unaccounted maternal and environmental effects may easily lead to their over-estimation (Merilä and Sheldon, 2001). Likewise, recent studies indicate that the strength of selection on phenotypic traits may commonly be overestimated due to environmental covariances between fitness and focal traits (Scheiner et al., 2002; Stinchcombe et al., 2002; Kruuk et al., 2003). In other words, environmental correlations between fitness and phenotypic traits are probably common since individuals in good condition may typically breed in habitats of best quality (e.g., Orians, 1969), and because environmental conditions affect fitness and many traits independently (Stinchcombe et al., 2002). This problem of incidental covariation, although recognised already a while ago (e.g., Price et al., 1988; Rausher, 1992), has been largely neglected in treatments of selection on phenotypic traits. Consequently, if the phenomenon is wide spread, also climate driven environmental change may be creating positive covariance among phenotypic traits and fitness, suggesting more selection on phenotypic traits than is actually occurring. This could explain why many phenotypic traits in natural bird populations have not evolved as predicted by Eq. (1) (review in Merilä et al., 2001a).

Maternal effects can, except from biasing estimates of quantitative genetic parameters, increase or decrease the expected response to selection depending on their correlations with other phenotypic traits (Wade, 1998). Maternal effects in themselves may also change with changes in environmental conditions, thereby affecting the rate of evolution in direct proportion to the change in the environment. For example, the size of eggs in the barn swallow is strongly affected by temperature during a period of several days that is preceding the actual laying date (Saino et al., 2004). In addition, egg contents of carotenoids—which act as free radical scavengers— and lysozyme, which is an anti-bacterial immune component of maternal origin, are both strongly affected by temperature during this period (Saino et al., 2004). Thus, egg size and quality can be strongly modified by weather conditions encountered by the mother, and such effects can have long lasting consequences for offspring performance and even performance of grand-offspring (Mousseau and Fox, 1998).

As to opposing influences of genetic and environmental effects masking an evolutionary change at the phenotypic level, Merilä et al. (2001b) recently demonstrated that the mean condition index of collared flycatchers has declined over the past two decades despite the fact that the trait was heritable and under consistent directional selection acting on both phenotypic and breeding values. Focusing on mean breeding value of condition over time, they found that it had increased over the same period, suggesting that the evolutionary response had actually taken place, but that it had been concealed by deteriorating environmental conditions over the same time period (Merilä et al., 2001b). This example illustrates the possibility that lack of change in the mean phenotype of a population over time does not need to imply lack of evolution at the genotypic level. This phenomenon, known as counter-gradient variation—and formally defined as the negative covariance among genotypic and environmental influences on a trait—is known to be common in the wild (Conover and Schultz, 1995). Consequently, caution should be exercised when drawing evolutionary inferences on the basis of phenotypes alone: patterns of change, or lack thereof, might look entirely different at the genotypic level (Conover and Schultz, 1995; Merilä et al., 2001a).

There are two additional genetic issues, which are worth mentioning in the context of climate change responses. First, since quantitative genetic parameters are affected by environmental conditions under which the traits are expressed, climate change may have a direct impact on the amount of genetic variation on which selection can act (Hoffmann and Parsons, 1997; Hoffmann and Merilä, 1999). Second, due to the climate-induced changes in population size (Møller et al., 2004, this volume) and structure (Berthold, 2001), gene frequencies and quantitative genetic parameters of populations might also be changing. We will briefly discuss implications of each of these issues in turn.

It is well established that quantitative trait parameters, such as additive genetic variances and heritabilities, may change depending on the environmental conditions under which they are measured (reviews in Møller and Swaddle, 1997; Hoffmann and Merilä, 1999; Merilä and Sheldon, 2001). However, it is less clear what kind of changes are to be expected under current climate change scenarios because the changes in the expression of additive genetic variance and heritability are not associated with any particular environmental conditions in any clear cut predictable manner (reviews in Hoffmann and Merilä, 1999; Merilä and Sheldon, 2001). Likewise, whether "new environments" created by climate change should be considered stressful or not is not directly obvious. Nevertheless, since climate change is believed to lead to increase in the variance in weather conditions (Easterling et al., 1997), this may mean that quantitative genetic parameters, and hence, predicted responses to selection, are also going to be more variable, if the amount of quantitative genetic variance depends on environmental conditions.

Many populations are predicted to decrease in size as a consequence of climate change (see Møller et al., 2004, this volume), and many populations of migratory species are predicted to change from migration to residency (see Böhning-Gaese and Lemoine, 2004, this volume), as already happened in many species during the last 200 years (Berthold, 2001). Residency is associated with a reduction in dispersal distance, and hence in a reduction in the level of mixing of neighbouring populations (Belliure et al., 2000). Smaller population sizes and reduced gene flow will most likely result in lowered effective population size, and consequently, in increased risk of losing genetic variation due to genetic drift (Frankham, 1999). Small population sizes will also increase the risk of inbreeding, and thereby also increased incidence of inbreeding depression which is well documented and typically severe in birds, e.g., in terms of hatching (e.g., Rowley *et al.*, 1993) or lifetime reproductive success (Kruuk *et al.*, 2002a,b and references therein). Further evidence comes from the observation that hatching success across a sample of 94 species of birds revealed a strong positive relationship between hatching failure and band sharing coefficients among "unrelated" adults (Spottiswoode and Møller, 2004). Effects of inbreeding directly related to the long-term effects of climate change may further affect the phenotypes of individuals in populations and, therefore, temporal trends in the mean phenotypes of populations. For instance, Schiegg *et al.* (2002) showed that inbred acorn woodpeckers *Melanerpes formicivorus* were unable to adjust their breeding time in response to climate change.

Adaptation to climate change may be constrained by the timing of migration. Both and Visser (2001) showed for the pied flycatcher breeding in the Netherlands that while climate change advanced the phenology of the breeding area, the timing of spring migration did not change sufficiently rapidly to accommodate this change in conditions at the breeding grounds. Such incongruence between optimal timing of migration and breeding is discussed in detail by Visser *et al.*, (2004, this volume) and Coppack and Pulido (2004, this volume).

C. Migration

Temporal change in the mean phenotype of a population could be attained also without phenotypic plasticity or genetic adaptation if new genotypes arrive to a population from other parts of species distribution range. Many bird species have changed their distribution range recently (Møller *et al.*, 2004, this volume), and consequently, it is possible that at least part of the changes in mean phenotypic traits over time might reflect effects of migration. In order to identify migration as a cause for change in the mean phenotype of a population, we need information about the extent of migration. This could be quantified either with direct methods (e.g., capture–mark–recapture studies) or with indirect methods relying on analyses of multi-locus genotypes (e.g., Gaggiotti *et al.*, 2001). At the moment, however, we are unaware of any studies showing that a temporal change in mean phenotype of a population could be attributable to migration (but see Böhning-Gaese and Lemoine, 2004, this volume).

Temporal change in phenotype of a population may not be due to evolutionary change in that particular population, but rather due to change in composition of the population due to gene flow from a neighbouring population. Different subspecies of birds often have different breeding and wintering grounds (Berthold, 2001). Thus, analyses of isotope profiles of such populations have provided evidence of divergence in two Scandinavian subspecies of the willow warbler *Phylloscopus trochilus* (Chamberlain *et al.*, 2001). More surprisingly, a recent study of a "population" of barn swallows breeding in Denmark revealed that the distribution of isotopes in feathers grown in the African winter quarters had clearly bimodal distribution (Møller and Hobson, 2004). These two different fractions of the population not only differed significantly in phenotype but also in phenotype of their offspring. Thus, adult birds breeding in sympatry may sometimes winter in very different areas and show dramatic differences in reproductive performance. Changes in climate may switch the balance in fitness advantage from one population to the other. Perhaps, with the exception of island populations such as those of Darwin's finches, very few populations are truly closed populations. Therefore, studies of phenotypic changes due to climate change should carefully examine the effects of phenotypic plasticity and micro-evolutionary change due to natural selection and due to gene flow.

V. CONCLUSIONS AND FURTHER DIRECTIONS

In our treatment, we have focused on how the patterns seen in analyses of long-term time series can only be understood when pattern and process are considered simultaneously. Thus, in order to understand causes and consequences of climate change on bird populations, we need not only high quality long-term data sets but also rigorous analytical protocols and methods to account for possible biases in the data. Likewise, our understanding of the mechanisms underlying observed changes (or lack thereof) is dependent on knowledge of genetics of the traits in question, as well as on experimental studies focusing on proximate determinants of trait variation within and among populations. It is only through careful experimentation that we will be able to make inferences about causal relationships and their relative importance. At the same time, experimental studies of birds in the context of climate change research are difficult, and the progress in the field is likely to proceed through combination of experimental and observational approaches. It is also worth emphasising the fact that a temporal change in the mean phenotype of a population may be constrained for several reasons, and these constraints may also change in the course of a long-term study. Therefore, studies of phenotypic changes due to climate change should carefully examine the effects of phenotypic plasticity, micro-evolutionary change and migration as potential explanations for observed changes.

Table 1 A check-list for analyses of long-term studies of avian populations

How representative is the study population for the species? Could local variation in habitat quality (or non-random dispersal) account for the observed variation in phenotype?

What is the sampling variance in the character of interest? Has this variance changed during the course of the study due to differences in intensity of study, differences in population size, or differences in procedures?

Is the climatic variable relevant for the study organism and the part of the annual cycle being investigated? For migratory species, have all relevant parts of the annual cycle been included in the analysis?

Has potential temporal auto-correlation in the data been accounted for, and have delayed (time-lagged) effects been tested for?

Have effects of density-dependence and delayed density-dependence been investigated?

Which are the potential mechanisms accounting for the temporal trend? Is mean phenotypic plasticity of a similar magnitude as the recorded change in mean phenotype?

Do differences in mean phenotype between cohorts, measured in units of standard deviations, correlate with the presumed climatic variable underlying the temporal trend?

Is the character heritable? Does the net intensity of selection estimated from the equation $R = h^2 S$ provide an estimate of selection similar to that calculated from comparisons of phenotypes before and after selection?

Have quantitative genetic parameters, such as heritability and breeding value, changed during the study period?

In Table 1, we have compiled a checklist that may provide crucial questions during the interpretation of analyses based on long-term data sets, both for the scientists involved in such studies and those attempting to interpret results of such studies. This checklist presents a summary of many of the issues that we have addressed in this chapter. We hope that this checklist will inspire not only empirical biologists involved in studies of the effects of climate change on birds and other organisms to make better research but also biologists and decision makers to be able to better interpret the available scientific literature. It is unlikely that any single study will be able to provide clear answers to all of these questions, but they may provide an important reference against which to judge the conclusion based on long-term studies.

ACKNOWLEDGEMENTS

P. Berthold, W. Fiedler and two reviewers kindly provided constructive comments on earlier version of this manuscript. J. M. has been supported by Academy of Finland.

REFERENCES

Aanes, R., Sæther, B.-E., Smith, F.M., Cooper, E.J., Wookey, P.A. and Øritsland, N. (2002) *Ecol. Lett.* **5**, 445–453.

Beckerman, A., Benton, T.G., Ranta, E., Kaitala, V. and Lundberg, P. (2002) *Trends Ecol. Evol.* **17**, 263–269.

Belliure, J., Sorci, G., Møller, A.P. and Clobert, J. (2000) *J. Evol. Biol.* **13**, 480–487.

Berthold, P. (2001) *Bird Migration.* Oxford University Press, Oxford.

Berthold, P., Mohr, G. and Querner, U. (1990) *J. Orn.* **131**, 33–45.

Böhning-Gaese, K. and Lemoine, N. (2004) (Ed. by A.P. Møller, W. Fiedler and P. Berthold), Birds and climate change. *Advances Ecol. Res.* **35**, 209–234.

Both, C. and Visser, M.M. (2001) *Nature* **411**, 296–298.

Box, G.E.P. and Jenkins, G.M. (1976) *Time Series Analysis: Forecasting and Control.* Holden-Day, San Francisco.

Burnham, K.P., Anderson, D.R., White, G.C., Brownie, C. and Pollock, K.H. (1987) *Am. Fish. Soc. Monogr.* **5**, Bethesda, MD.

Chamberlain, C.P., Bensch, S., Feng, X., Åkesson, S. and Andersson, T. (2001) *Proc. R. Soc. Lond. B* **267**, 43–48.

Cooch, E.G. (2002) *J. Appl. Stat.* **29**, 143–162.

Conover, D.O. and Schultz, E.T. (1995) *Trends Ecol. Evol.* **10**, 248–252.

Coppack, T. and Pulido, F. (2004) (Ed. by A.P. Møller, W. Fiedler and P. Berthold), Birds and climate change. *Advances Ecol. Res.* **35**, 129–148.

Easterling, D.A., Horton, B., Jones, P.D., Peterson, T.C., Karl, T.R., Parke, D.E., Salinger, M.J., Razuvayev, M., Plummer, N., Jameson, P. and Folland, C.K. (1997) *Science* **277**, 364–367.

Forchammer, M., Post, E. and Stenseth, N.-C. (2002) *J. Anim. Ecol.* **71**, 1002–1014.

Frankham, R. (1999) *Genet. Res.* **74**, 237–244.

Gaggiotti, O.E., Kones, F., Lee, W.M., Amos, W., Harwood, J. and Nichols, R.A. (2001) *Nature* **416**, 424–427.

Grant, P.R. (1986) *Ecology and Evolution of Darwin's Finches.* Princeton University Press, Princeton.

Grant, B.R. and Grant, P.R. (1989) *Evolutionary Dynamics of a Natural Population.* University of Chicago Press, Chicago.

Hoffmann, A.A. and Merilä, J. (1999) *Trends Ecol. Evol.* **14**, 96–101.

Hoffmann, A.A. and Parsons, P.A. (1997) *Extreme Environmental Change and Evolution.* Cambridge University Press, Cambridge.

Hurrell, J.W. (1995) *Science* **269**, 676–679.

Kruuk, L.E.B., Sheldon, B.C. and Merilä, J. (2001) *Am. Nat.* **158**, 557–571.

Kruuk, L.E.B., Sheldon, B.C. and Merilä, J. (2002a) *Proc. R. Soc. Lond. B* **150**, 1581–1589.

Kruuk, L.E.B., Slate, J., Pemberton, J.M., Brotherstone, S., Guinness, F. and Clutton-Brock, T.H. (2002b) *Evolution* **56**, 1683–1695.

Kruuk, L.E.B., Merilä, J. and Sheldon, B.C. (2003) *Trends Ecol. Evol.* **18**, 207–209.

Larsson, K., van der Jeugd, H.P., van der Veen, I.T. and Forslund, P. (1998) *Evolution* **52**, 1169–1184.

Lebreton, J.-D., Burnham, K.P., Clobert, J. and Anderson, D.R. (1992) *Ecol. Monogr.* **62**, 67–118.

Lehikoinen, E., Sparks, T.H. and Zalakevicius, M. (2004) (Ed. by A.P. Møller, W. Fiedler and P. Berthold), Birds and Climate Change, *Advances Ecol. Res.* **35**, 1–30.

Lindström, J. (1999) *Trends Ecol. Evol.* **14**, 343–348.

Lynch, M. and Walsh, B. (1998) *Genetics and Analysis of Quantitative Traits.* Sinauer, Sunderland.

Mason, C.F. (1995) *Bird Study* **42**, 182–189.

Merilä, J. and Sheldon, B.C. (2001) *Curr. Ornithol.* **16**, 179–255.

Merilä, J., Kruuk, L.E.B. and Sheldon, B.C. (2001a) *Genetica* **112/113**, 199–222.

Merilä, J., Kruuk, L.E.B. and Sheldon, B.C. (2001b) *Nature* **412**, 76–79.

Metcalfe, N.B. and Monaghan, P. (2001) *Trends Ecol. Evol.* **16**, 254–260.

Monros, J.S., Belda, E.J. and Barba, E. (2002) *Oikos* **99**, 481–488.

Møller, A.P. (1994) *Behav. Ecol. Sociobiol.* **35**, 115–122.

Møller, A.P. (2002) *J. Anim. Ecol.* **71**, 201–210.

Møller, A.P. (2004) Global Change Biol. (in press).

Møller, A.P. (2004) Unpublished manuscript.

Møller, A.P. and Hobson, K.A. (2004) *Proc. R. Soc. Lond. B* **271**, 1355–1362.

Møller, A.P. and Swaddle, J.P. (1997) *Asymmetry, Developmental Stability, and Evolution.* Oxford University Press, Oxford.

Møller, A.P. and Szép, T. (2004) J. Evol. Biol. (in press).

Møller, A.P., Berthold, P. and Fiedler, W. (2004) (Ed. by A.P. Møller, W. Fiedler and P. Berthold), Birds and Climate Change, *Advances Ecol. Res.* **35**, 235–243.

Mousseau, T.A. and Fox, C.W. (Eds.) (1998) *Maternal Effects.* Oxford University Press, New York.

Orians, G.H. (1969) *Am. Nat.* **103**, 589–603.

Palmer, M.R. and Pearson, P.N. (2003) *Science* **300**, 480–482.

Parmesan, C. and Yohe, G. (2003) *Nature* **421**, 37–42.

Price, T., Kirkpatrick, M. and Arnold, S. (1988) *Science* **240**, 798–799.

Prince, S.D. and Justice, C.O. (Eds.) (1991) *Int. J. Remote Sensing* **12**, 1133–1421.

Przybylo, R., Sheldon, B.C. and Merilä, J. (2000) *J. Anim. Ecol.* **69**, 395–403.

Pulido, F., Berthold, P. and van Noordwijk, A.J. (1996) *Proc. Natl. Acad. Sci. USA* **93**, 14642–14647.

Pulido, F., Berthold, P., Mohr, G. and Querner, U. (2001) *Proc. R. Soc. Lond. B* **268**, 953–959.

Pulido, F. and Berthold, P. (2004) (Ed. by A.P. Møller, W. Fiedler and P. Berthold), Birds and Climate Change, *Advances Ecol. Res.* **35**, 149–181.

Rausher, M.D. (1992) *Evolution* **46**, 616–626.

Root, T.L., Price, J.L., Hall, K.R., Schneider, S.H., Rosenzweig, C. and Pounds, A.J. (2003) *Nature* **421**, 57–60.

Rossiter, M.C. (1996) *Annu. Rev. Ecol. Syst.* **27**, 451–476.

Rowley, I., Russell, E. and Brooker, M. (1993) In: *The Natural History of Inbreeding and Outbreeding* (Ed. by N.W. Thornhill), pp. 304–328. University of Chicago Press, Chicago.

Sæther, B.-E., Sutherland, W.J. and Engen, S. (2004) (Ed. by A.P. Møller, W. Fiedler and P. Berthold), Birds and Climate Change, *Advances Ecol. Res.* **35**, 183–207.

Saino, N., Romano, M., Ambrosini, R., Ferrari, R.P. and Møller, A.P. (2004) *Funct. Ecol.* **18**, 50–57.

Sanz, J.J. (2002) *Global Change Biol.* **8**, 409–422.

Sanz, J.J. (2003) *Ecography* **26**, 45–50.

Scheiner, S.M., Donohue, K., Dorn, L.A., Mazer, S.J. and Wolfe, L.M. (2002) *Evolution* **56**, 2156–2167.

Schiegg, S., Pasinelli, G., Jeffrey, R., Walters, J.R. and Daniels, S.J. (2002) *Proc. R. Soc. Lond. B* **269**, 1153–1159.

Sheldon, B.C., Kruuk, L.E.B. and Merilä, J. (2003) *Evolution* **57**, 406–420.

Sillett, T.S., Holmes, R.T. and Sherry, T.W. (2000) *Science* **288**, 2040–2042.

Spottiswoode, C. and Møller, A.P. (2004) *Proc. R. Soc. Lond. B* **271**, 267–272.

Stinchcombe, J.R., Rutine, M.T., Burdick, D.S., Tiffin, P., Rausher, M.D. and Mauricio, R. (2002) *Am. Nat.* **160**, 511–523.

Tryjanowski, P. and Sparks, T. (2001) *Int. J. Biometeorology* **45**, 217–219.

Visser, M.E., Adriaensen, F., van Balen, J.H., Blondel, J., Dhondt, A.A., van Dongen, S., du Feu, C., Ivankina, E.V., Kerimov, A.B., de Laet, J., Matthysen, E., McCleery, R., Orell, M. and Thomson, D.L. (2003) *Proc. R. Soc. Lond. B* **270**, 367–372.

Visser, M.E., Both, C. and Lambrechts, M.M. (2004) (Ed. by A.P. Møller, W. Fiedler and P. Berthold), Birds and Climate Change, *Advances Ecol. Res.* **35**, 87–108.

Wade, M.J. (1998) In: *Maternal Effects as Adaptations* (Ed. by T.A. Mousseau and C.W. Fox), pp. 5–21. Oxford University Press, New York.

Walther, G.R., Post, E., Convey, P., Menzel, A., Parmesan, C., Beebee, T.J.C., Fromentin, J.-C., Hoegh-Guldberg, O. and Bairlein, F. (2002) *Nature* **416**, 389–395.

White, G.C. and Burnham, K.P. (1999) *Bird Study* **46**, S120–S138.

Photoperiodic Response and the Adaptability of Avian Life Cycles to Environmental Change

TIMOTHY COPPACK* AND FRANCISCO PULIDO

I. SUMMARY

In birds, the annual change in daylength is the most important environmental cue used for synchronising breeding, moult, and migration with recurrent seasonal fluctuation in environmental conditions. Human-caused environmental changes may affect photo-responsive birds in two ways: (1) The photoperiod may become an unreliable predictor of favourable conditions if the phase relationship between temperature-dependent resource availability and daylength changes. For example, advances in the timing of breeding in response to increased spring temperature expose juvenile birds to altered photoperiodic conditions, which may result in unseasonably early autumn migration. (2) Range shifts and expansions may expose birds to novel photoperiodic conditions. Extant responses to these conditions could limit the potential of birds to evade increasingly unsuitable habitats and to establish new breeding and wintering

---

E-mail address: coppack@web.de (T. Coppack)

ADVANCES IN ECOLOGICAL RESEARCH, VOL. 35
0065-2504/04 $35.00 DOI 10.1016/S0065-2504(04)35007-5

grounds. However, if birds respond to novel photoperiodic conditions in an adaptive way—i.e., the elicited phenotypic change is in accord with the direction of selection—then adaptation of avian life cycles to global environmental change will be facilitated. In the course of environmental change, we expect the photoperiodic response itself to be the target of selection. However, adaptive evolution of the response to daylength may not keep pace with rapid environmental changes because of unfavourable genetic correlations among life-cycle stages or the lack of within-population genetic variation in phenotypic plasticity.

II. INTRODUCTION

A. General Introduction

The Earth's atmosphere is currently warming at an unprecedented rate (see "Preface"). This is expected—and has been shown—to have profound ecological and evolutionary consequences (McCarty, 2001; Walther et al., 2002). The best documented biological response to recent climatic change is the change in the timing of seasonal events in plants, insects, and birds (Parmesan and Yohe, 2003; Root et al., 2003). Changes in ambient temperature may affect animal and plant phenology directly or may have indirect effects by changing the environment in which selection occurs, i.e., by altering the availability of food resources and the prevalence of predators, parasites, competitors, and mates. Species that rely on the annual change in daylength for timing their life cycles, like most birds, will become desynchronised with the temperature-dependent selective environment, if photoperiodic responses do not change correspondingly (Visser et al., 1998, 2004, this volume; Visser and Holleman, 2001).

A dramatic rise in global temperature will not only alter the timing of seasons but also the geographic location of suitable habitats, which, in turn, will have an impact on the distribution of animal and plant species—in particular highly mobile organisms, like birds (Root, 1993; Thomas and Lennon, 1999). Whether range shifts and colonisation of new habitats are possible hinges upon the ability of species and populations to appropriately respond to novel photoperiodic conditions. If the new conditions have rarely or never been encountered before in the evolutionary history of the population, adaptive reaction norms may not exist. De novo evolution of phenotypic plasticity and the evolvability of reaction norms may be constrained by the lack of genetic variation or by unfavourable genetic correlations with other life-history traits. Alternatively, if global environmental changes have recurrently been experienced, birds may have evolved highly adjustable photoperiodic response systems that adapt them to a wide range of environmental conditions.

The aim of this chapter is to give an overview of the consequences of photoperiodic responsiveness for the adaptability of birds to human-caused environmental change. It will focus on the question to what extent photoperiodic responses facilitate or constrain the adaptation of avian life cycles to novel environmental conditions.

B. Photoperiodic Control of Avian Seasonality

Birds match their reproduction with periods of highest resource availability (Lack, 1968; Perrins, 1970). The ultimate cause for this close synchronisation between breeding and the peak in food abundance is the high resource demand during chick-rearing (e.g., Daan *et al.*, 1988; van Noordwijk *et al.*, 1995; Nager *et al.*, 1997; Thomas *et al.*, 2001). In contrast to mammals, most birds are not capable of storing nutrients for raising their young and depend on specific protein-rich food resources, which are often only available during a short period. In principle, food abundance itself could function as the positive stimulus for initiating egg laying (Daan *et al.*, 1988; Nager *et al.*, 1997). However, in seasonal environments, initial predictive cues (*sensu* Wingfield, 1980) are required for initiating spring migration and gonadal development well in advance of favourable breeding conditions.

The seasonal change in daylength provides the most reliable source of temporal information about the environment and has been adopted by birds in the course of evolution as the main environmental cue for synchronising reproduction, moult, and migration with favourable environmental conditions. Other factors, such as temperature, rainfall, food abundance, and social stimuli may also affect the timing of life-cycle stages, but provide only short-term predictive information. These factors serve as supplementary cues for fine-tuning the rate of gonadal growth and the timing of breeding with local phenological conditions (see reviews by Wingfield *et al.*, 1992, 1993; Hahn *et al.*, 1997; Visser and Lambrechts, 1999).

Many empirical studies have explored the role of daylength in the control of avian seasonality (for reviews see Dawson *et al.*, 2001; Dawson, 2002). Most studies have been carried out with resident or short-distance migratory species of the temperate zone. Outside the tropics, the time of year is immediately apparent from the regular sinusoidal change in daylength, and all species experiencing these conditions appear to be photoperiodic. Birds living in tropical regions, in contrast, experience only slight changes in photoperiod and have been considered to be unable to use daylength as a cue for controlling seasonality (e.g., Dittami and Gwinner, 1985). However, recent evidence suggests that even

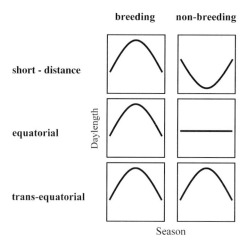

Figure 1 Photoperiodic conditions experienced on the breeding and non-breeding grounds by different types of migrants (schematic).

tropical species have evolved (or retained) the ability to respond to photoperiodic change (Hau, 2001; Styrsky *et al.*, 2004).

In migratory songbirds, which encounter complex photoperiodic conditions throughout the year (Figure 1), daylength acts as a synchroniser (*Zeitgeber*), entraining endogenous circannual rhythms of gonadal maturation, moult, and nocturnal migratory activity (Gwinner, 1986, 1996a,b; Berthold, 1996). With regard to the photoperiodic response, the annual cycle of a migratory bird can be divided into two phases in which short and long days exert different effects on the timing of life-cycle events (cf., Gwinner, 1996b).

During the post-breeding phase, from summer to late autumn, the onset of juvenile/post-breeding moult is advanced and its duration reduced by short photoperiods. Consequently, autumn migratory activity is initiated at an earlier age. This accelerated transition of life-cycle stages is supposed to be adaptive in that it enables birds hatching or breeding late in the season to moult and initiate migration before conditions deteriorate on the breeding grounds (Berthold, 1996). Long days after breeding induce gonadal regression and reproductive photorefractoriness, i.e., insensitivity of the hypothalamus–pituitary–gonadal axes towards long-day stimulation (reviewed by Hahn *et al.*, 1997; Nicholls *et al.*, 1988; Dawson *et al.*, 2001). Sexual immaturity after hatching may be caused by an innate photorefractory state, similar to photorefractoriness in adult birds, and exposure to short days generally promotes the termination of photorefractoriness. In the European starling (*Sturnus vulgaris*), juvenile photorefractoriness is terminated by short photoperiods, and premature exposure to short days leads to early gonadal development (Williams *et al.*, 1987). In the garden warbler (*Sylvia borin*)—a Palaearctic long-distance migrant that winters in central and southern Africa (Moreau, 1972)—termination of photorefractoriness and the post-refractory increase in photosensitive is predominantly controlled by endogenous factors (Gwinner *et al.*, 1988), but is possibly also modified by daylength conditions.

In the prenuptial phase, i.e., from winter through spring, the initiation of spring migration and gonadal growth are advanced by long (or increasing) daylengths (e.g., Rowan, 1925; for an example of such a response in a non-passerine migrant, see Rees, 1982, 1989). Also in long-distance migratory species wintering close to the Equator or beyond it (e.g., garden warbler, spotted flycatcher, *Muscicapa striata*), spring departure from the wintering site may strongly depend on photoperiodic conditions (e.g., Gwinner, 1987; Kok *et al.*, 1991). Experimental data suggest that garden warblers overwintering south of the Equator—where they experience longer daylengths than conspecifics wintering further north—advance prenuptial moult, gonadal growth, and the onset of spring migratory activity in response to the photoperiodic environment (Gwinner, 1987). It is supposed that this advancement of spring migratory disposition is adaptive in that it enables individuals wintering far south to reach their northern breeding grounds in time, in spite of the longer distance they have to travel (Gwinner, 1987, 1996a).

An additional point must be made about the significance of photoperiodic responses for birds breeding (or hatching) at very high latitude, i.e., under continuous light. Experimental data on captive bluethroats (*Luscinia svecica*) breeding at 66°N latitude show that the timing of post-breeding moult is inflexible against photoperiodically imposed time stress and suggest that the timing of moult is set early in the season and is endogenously determined (Lindström *et al.*, 1994). However, experiments by Pohl (1999) imply that passerine birds breeding in high Arctic regions may well use daily changes in the spectral composition of sunlight as a cue for synchronising physiological and behavioural rhythms, if periodic changes in light intensity are unperceivable.

C. Photoperiodic Response and Environmental Change

Climate change has been shown to cause a decoupling of phenological relationships between predators and their prey (Buse *et al.*, 1999; Visser and Holleman, 2001; Stenseth and Mysterud, 2002). For example, the timing of the great tit's (*Parus major*) breeding season in the Netherlands has remained unaffected by increasing spring temperatures, even though the date of highest caterpillar abundance has become earlier and selection for earlier breeding has intensified (Visser *et al.*, 1998). In this population, an adaptive advancement of breeding has not occurred, most likely because the cues used for initiating reproduction have not changed with the advancement of spring (Visser *et al.*, 1998). It is likely that this mismatch between resource availability and demand is in part the result the species' rigid photoperiodic control system. Great tits depend strongly on photoperiodic stimuli for initiating gonadal growth (Silverin, 1994), and intraspecific variation in egg-laying date is attributable to among-population differences in photorespon-siveness (Figure 2; Silverin *et al.*, 1993). Similarly, differences in the timing of breeding between blue tit (*P. caeruleus*) populations adapted to different seasonal environments (southern France versus Corsica) are

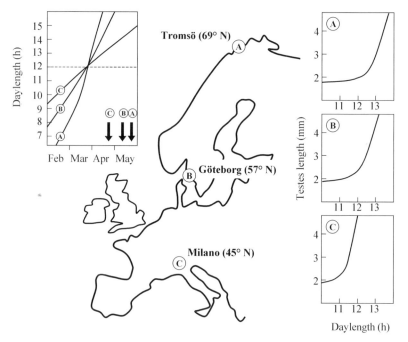

Figure 2 Results from a common garden experiment on male great tits from three different European populations (adapted from Silverin *et al.*, 1993). Birds were collected in winter and transferred to Göteborg (Sweden) where they were kept on short days (8L:16D). All birds were subsequently exposed to an artificial increase in daylength (0.5 h per week) beginning in early January. Diagrams on the right approximate the induced testicular growth in relation to daylength (12 h of light corresponds to week 8 of the experiment). The left diagram shows the changes in daylength the birds would have normally experienced on their respective breeding grounds. The bold arrows show the mean egg-laying dates for each population.

controlled by differences in the responsiveness to photoperiodic stimulation (Lambrechts *et al.*, 1996, 1997). Corresponding results have been obtained on a smaller geographical scale for two adjacent blue tit populations on Corsica (Blondel *et al.*, 1999; Lambrechts *et al.*, 1999). In general, birds which rely on a single, rigid response mechanism for timing reproduction may be unable to immediately track changes in the seasonal availability of food.

Besides phenological changes, range shifts are the most probable and best documented response of birds to global climate change (Walther *et al.*, 2002). Climatic amelioration during the last decades has resulted in a northward expansion of breeding ranges (Burton, 1995; Thomas and Lennon, 1999; Brommer, 2004). At the same time, the wintering areas of some migratory species have shifted northwards (reviewed by Burton, 1995; Berthold, 1998; Fiedler, 2003; Valiela and Bower, 2003). Range shifts are caused by changes in the geographical distribution of suitable climatic conditions and concomitant shifts in resource availability (Root, 1988a,b, 1993). Species follow the environmental conditions they are adapted to and evade areas, which are no longer optimal. Latitudinal range shifts involve changes in photoperiodic conditions. Poleward shifts increase the amplitude of seasonal daylength changes. Birds are bound to respond to these changes.

Most species breeding at mid to high latitude have higher photoperiodic response thresholds than more southerly distributed species, causing a later onset of gonadal recrudescence with increasing latitude. For instance, in European great tits (Figure 2) and various North American *Zonotrichia* species and sub-species the level of sensitivity towards long days is negatively correlated with the species' distributional range, being lowest in the species with the most northerly distribution (Miller, 1960; Lofts and Murton, 1968; Silverin *et al.*, 1993). If global temperature suddenly rises, and if the optimal time for breeding advances, adaptation of the timing of breeding will be most severely constrained in northern temperate species with relatively high photoperiodic response thresholds. On the other hand, if birds of southern origin (e.g., tropical birds) with relatively low thresholds extend their ranges to the north, they will become exposed to steeper vernal increases in daylength, which could result in unseasonably early gonadal development and egg laying. Moreover, longer days encountered at higher latitude in summer could lead to an unseasonably early onset of gonadal regression and photorefractoriness, which would be maladaptive under conditions favouring an extended reproductive period and multiple broods. However, if novel photoperiodic condition bring forth changes in annual rhythmicity that are adaptive under altered climatic conditions, adaptation to environmental change may be facilitated, or even reinforced. The existence of highly adjustable photoperiodic responses that adapt birds to a wide range of different environments is implied by the successful introduction of some European bird species to New Zealand (Cockrem, 1995).

III. RESPONSES TO TEMPORAL AND SPATIAL VARIATION IN PHOTOPERIODIC CONDITIONS

A. Photoperiodic Response of Migrants to Changes in Hatching Date

Spring temperatures in northern temperate regions have considerably increased over the last century (Sparks and Menzel, 2002). In response, numerous bird species are laying their eggs earlier in the year (e.g., Crick and Sparks, 1999; reviewed by Coppack and Both, 2002 and Walther et al., 2002). Earlier egg laying leads inevitably to earlier hatching, since the time for an incubated egg to develop is relatively fixed (but see Cresswell and McCleary, 2003). Unlike birds hatched in midsummer, birds hatching earlier in the year are exposed to short and increasing daylengths during their first days of life and experience the summer solstice at an older age. How do these altered photoperiodic conditions affect the timing of juvenile moult and autumn migration?

The response of the age at onset of autumn migratory activity to differences in hatching date, i.e., photoperiodic conditions, can be described as a reaction norm (Pulido, 2000; Pulido et al., 2001a). Within the range of hatching dates currently found in southern German blackcaps (*Sylvia atricapilla*), the population reaction norm is close to linear (Figure 3). The later a bird hatches, the younger it is when it completes juvenile moult and initiates autumn migratory activity. Acceleration of juvenile development with later hatching is caused by short (or decreasing) daylengths ("calendar effect", Berthold, 1996). Birds hatched earlier in the season experience long daylengths for a longer period, and this causes autumn migration to commence at an older age. However, differences in hatching date are only partially compensated: for every 2 days a blackcap hatches earlier, it is about 1 day older when it initiates migration (Pulido, 2000).

When blackcaps from the southern German breeding population were kept under artificial daylengths simulating 6 weeks earlier hatching, moult ended and migration was initiated at an older age than under natural photoperiodic conditions (Figure 3; Coppack et al., 2001). However, this response was not strong enough to compensate for the simulated advancement of hatching date. The results from this experiment suggest that earlier hatching will inevitably lead to earlier autumn migration. An earlier onset of migration with earlier breeding is likely to be maladaptive in a persistently warming environment (but see Jenni and Kéry, 2003 for the potential advantage of earlier migration in trans-Saharan migrants). Global warming is expected to reduce the probability of cold spells in autumn. Under these conditions, migrants could afford to stay longer on the

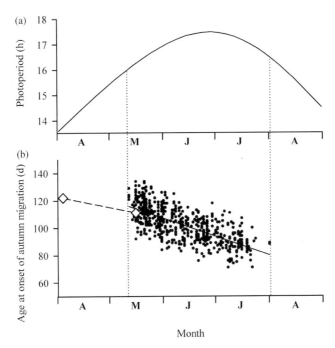

Figure 3 Relationship between hatching date and age at onset of autumn migratory activity as measured under laboratory conditions in southern German blackcaps (*S. atricapilla*) reared in 1988–1997 under natural photoperiodic conditions (top diagram: daylength conditions at 47°50′N latitude). The linear regression (*solid line*) gives the mean population reaction norm (adapted from Pulido, 2000). *Vertical dotted lines* indicate the range of hatching dates under study. The *open symbols* represent the mean onset of autumn migratory activity as measured under simulated early and natural hatching dates in a split brood-experiment (see text and Coppack *et al.*, 2001 for details). The connecting *dashed line* gives the reaction norm outside the current range of hatching dates.

breeding grounds. Later departure in autumn would then allow first-year birds to explore and become familiar with future breeding sites, and to take full advantage of specific food resources used for premigratory fattening—provided food is still available and interspecific competition does not increase. Because photoperiodic responses to conditions outside the current range of hatching dates do not lead to a delayed migration date, extant reaction norms could reduce the fitness benefits of earlier breeding and, as a consequence, slow down life-cycle adaptation to global warming.

In addition, a close correlation between moult and migration (Pulido and Coppack, 2004) might set limits to the independent evolution of breeding and migration phenology. Assuming that current climatic changes favour earlier

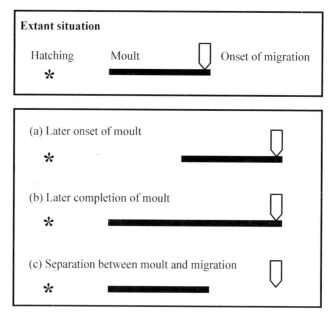

Figure 4 Schematic representation of three possible mechanisms leading to an extension of the time interval between hatching and the onset of autumn migration in juvenile songbirds. They may achieve delayed migration by either (a) postponing the onset of moult, by (b) delaying the completion of moult, causing a prolongation of moult duration, or by (c) increasing the temporal separation of moult and migration. In the presence of high genetic correlation between moult–migration overlap and the onset of migration, as found in the blackcap (Pulido and Coppack, 2004), evolution of delayed onset of migration by mechanisms (b) and (c) may be constrained. If moult and migration are independent processes—or in case of sufficient genetic variation in the phase relationship between moult and migration—a delayed onset of migration could be accomplished by mechanism (c).

breeding and later autumn migration, there are three possible ways birds could change the timing of juvenile moult and the onset of migration relative to hatching date (Figure 4). The first would be to lengthen the interval between hatching and moult. The second potential mechanism would be to increase the duration of moult. Finally, the interval between hatching and onset of migration could be prolonged by extending the time interval between the termination of moult and the onset of migration. Results from a first study investigating the potential for adaptive changes in the timing of moult and migration suggest that in the blackcap, adaptation of the annual cycle may be constrained by unfavourable genetic correlations between the termination of moult and the onset of migration (Pulido and Coppack, 2004). Evolutionary changes in the termination of moult may not lead to adaptive changes in the onset of migration because, as a correlated selection response, the overlap between moult and migration will also be affected. The strong response to artificial selection for older age at onset of migration found in south German blackcaps (Pulido *et al.*, 2001b) was probably achieved by a delayed onset of moult (Figure 4, mechanism a). Significant heritabilities of the age at onset of juvenile moult in the blackcap and other species (Helm and Gwinner, 1999, 2001; Widmer, 1999) suggest that evolutionary change by this mechanism is possible.

The study on the relationship between hatching date and timing of migration in the blackcap illustrates the major problems we face when investigating the evolvability of reaction norms: what are the mechanisms underlying reaction norms? What is the prime target of selection? What determines phenotypic and genetic variability of reaction norms? Generally, we expect evolutionary change in reaction norms to be slow, as empirical and theoretical studies have found that reaction norms have low heritabilities and respond weakly to natural selection (cf. Pigliucci, 2001). However, despite these findings indicating low evolvability, small-scale differentiation in phenotypic plasticity has been found for a number of traits and in different organisms, including birds. In central European garden warblers, for instance, altitudinal variation in the timing of autumn migration is attributable to genetic differences in photoperiodic responsiveness. Population reaction norms appear to be adaptive, reflecting differences in the reliability of daylength as a predictive cue for seasonal changes in food availability in different habitats. Similarly, European blackbirds (*Turdus merula*) from an urban and an adjacent forest population clearly differed in the response of moult onset and duration to variation in hatching date in a common environment, suggesting genetic differences in reaction norms between populations (Partecke, 2002). Variation in the timing and duration of juvenile moult in response to the photoperiod has also been found among sub-species of the stonechat (*Saxicola torquata*) (Helm and Gwinner, 1999, 2001).

If climatic changes persist we expect evolutionary changes in reaction norms, because environmental cues used for synchronising the timing of breeding with maximal food availability may become unreliable (Visser *et al.*, 1998; Both and Visser, 2001; Sanz *et al.*, 2003). However, despite accumulating evidence that photoperiodic responses may differ between adjacent bird populations (Section II), we currently do not know to what extent adaptive changes in reaction norms will keep track with rapid environmental changes. A first quantitative genetic analysis of phenotypic plasticity in the timing of autumn migration in response to hatching date in the blackcap suggests that even though additive genetic variance is present for this trait (Pulido *et al.*, 2001a; Pulido and Berthold, 2003), selection responses may be too weak to allow rapid changes of the population mean.

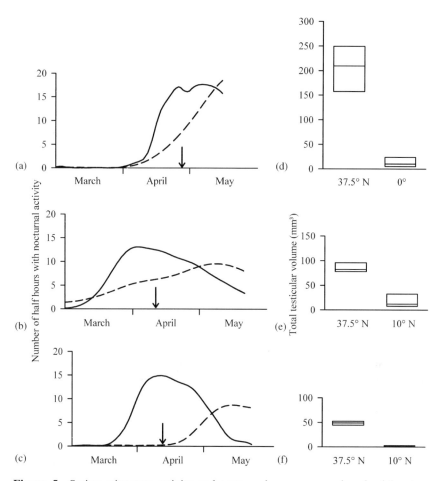

Figure 5 Spring migratory activity and testes volumes measured under laboratory conditions in yearling male garden warblers (a, d), common redstarts (b, e), and pied flycatchers (c, f). 14 garden warblers, 15 common redstarts, and 15 pied flycatchers from south Germany were hand-raised and kept for one year in captivity. At 40 days of age, birds were separated into two groups (treatment and control) in which each family was represented with at least one nest mate. Until the beginning of autumn migration in mid-August, birds of both groups were held under the photoperiodic conditions of their natal area. The treatment group was subsequently exposed to daylengths mimicking migration to and wintering in southern Europe at 37.5°N latitude. The control group was exposed to photoperiodic conditions approximating those experienced in their respective winter ranges in Africa, i.e., at the Equator (garden warbler) and at 10°N (common redstart; pied flycatcher). Light intensity was kept constant at 400 lx during the day, and at 0.01 lx at night. Solid lines in a–c give the conditional mean of nocturnal migratory activity of birds kept on the 37.5°N simulation, dashed lines represent the mean of control birds on African photoperiods. Lines in a–c are locally weighted within optimal bandwidths using the

B. Photoperiodic Response of Migrants to Novel Wintering Ranges

Ever since Rowan's (1925) fundamental discovery that long days in winter induce a change in physiological state (which in turn induces migration to the breeding grounds), researchers have asked whether migratory species have adapted their photoperiodic response mechanism to the diverse photoperiodic conditions experienced during equatorial and trans-equatorial migration (e.g., Hamner and Stocking, 1970; Gwinner, 1987, 1989). In fact, there are only a few experimental studies approaching this question, and even less is known about responses to photoperiodic conditions outside extant breeding and wintering ranges. This information, however, is decisive for predicting whether adaptive changes in migratory behaviour in response to an advancement of spring on the breeding grounds are possible (Coppack and Both, 2002).

If Palaearctic long-distance migrants were able to winter in Mediterranean regions instead of central or southern Africa, their migration routes would be considerably shorter which would allow them to arrive earlier on the breeding grounds and to breed earlier. In fact, the number of Palaearctic-African migrants observed in Mediterranean regions in winter has increased over the last few decades (Berthold, 2001). However, at higher latitudes, Palaearctic-African migrants encounter photoperiodic conditions that are markedly different from those experienced on their traditional African wintering grounds: daylengths decrease more rapidly in autumn, become considerably shorter in winter, and increase at a higher rate in spring (Figure 1). These conditions could potentially impede adaptive changes of the annual cycle. On the other hand, exposure to relatively short days during winter could terminate photorefractoriness at an earlier stage, which in turn would lead to more pronounced responses to the vernal increase in daylength and an advancement of springtime activities. This response to the novel photoperiodic environment would be advantageous under conditions favouring earlier spring arrival and breeding.

The photoperiodic effects of northern wintering has been recently tested in three species of Palaearctic-African passerine migrants—the garden warbler, the common redstart (*Phoenicurus phoenicurus*), and the pied flycatcher (*Ficedula hypoleuca*) (Coppack *et al.*, 2003). Male birds were exposed to photoperiodic conditions simulating wintering north of the Sahara desert at 37.5°N latitude. The three species of migrants responded to the experimental change in wintering area by initiating spring migratory activity earlier and by advancing testicular growth (Figure 5). The northern photoperiodic regime elicited the strongest

---

generalised-cross-validation method. Nocturnal (migratory) activity is given in number of half hours with activity. Arrows indicate the dates at which testes volumes were measured non-invasively in birds of both groups. Total testicular volumes are given as Box-plots in d–f (adapted from Coppack *et al.*, 2003).

advancement of spring migratory activity in the pied flycatcher. In all three species, testicular volumes in April—i.e., the time when these species usually arrive on the breeding grounds—were significantly larger in individuals exposed to Mediterranean photoperiods than in nest mates kept under African light-dark cycles. Again, the effect was strongest in the pied flycatcher (Coppack *et al.*, 2003).

These findings are in general agreement with results obtained in a Nearctic trans-equatorial migrant, the bobolink (*Dolichonyx oryzivorus*) (Engels, 1959, 1961). This species breeds in North America and normally stays from November through March in South America, south of the Equator. Male bobolinks kept in an outdoor aviary at 36°N latitude developed beak pigmentation (indicative of gonadal recrudescence) approximately in phase with all-year-round resident species (Engels, 1961). In earlier experiments on bobolinks, Engels (1959) had found that the rate of response to long photoperiods depended inversely on the length of the photoperiod to which males had been exposed previously during the post-nuptial photorefractory phase. Engels (1961) was to conclude that the bobolink could establish a wintering population north of the Equator without readjustment of its photoperiodic control system. However, the precise physiological mechanism underlying this flexible response to daylength remains unknown.

An advancement of life-cycle events with shorter migration distance is in accord with the expected selection response to earlier spring conditions. We therefore expect photoperiodic responses along the latitudinal gradient to reinforce selection for shorter migration distance if temperatures on the breeding grounds continue to rise. Although it is possible that the strength of the photoperiodic response to northern latitudes is maladaptive under the prevailing climatic conditions—as individuals wintering further north may arrive too early on the breeding grounds—we believe that long-distance migrants will gain high fitness benefits from wintering further north and advancing springtime activities under warmer climatic conditions. Similar conclusions have been drawn from blackcaps that have established a new wintering area in the British Isles (Terrill and Berthold, 1990). Over the last decades, an increasing number of blackcaps breeding in continental Europe is overwintering in Britain and Ireland (Langslow, 2002). Birds that winter in Britain rather than in southern Europe are exposed to photoperiodic conditions that promote an earlier termination of photorefractoriness and cause an accelerated onset of vernal physiological condition (Terrill and Berthold, 1990). This response to the novel photoperiodic environment in combination with higher winter survival of blackcaps feeding at bird tables may have facilitated rapid evolution of a novel migration strategy (Terrill and Berthold, 1990; Berthold *et al.*, 1992).

An additional point must be made about the possibility that global warming may cause Palaearctic-African migrants to make even longer journeys than today. At the same time as breeding grounds expand to the north (Thomas and Lennon, 1999), suitable wintering habitats may shift further south due to increasing aridity in the Sahel zone (e.g., Ottosson *et al.*, 2002). Birds may follow suitable breeding and wintering conditions, and by doing so, they will encounter longer and longer photoperiods on either side of the Equator. For some species, however, an expansion of the wintering grounds to the south may be impossible, because daylengths experienced close to and south of the Equator may be too long to permit termination of photorefractoriness. In south German pied flycatchers, for instance, normal annual rhythmicity was developed only under daylengths simulating wintering at 10°N latitude; under photoperiodic condition mimicking migration to wintering areas at 0° and 20°S latitude, prenuptial moult was disrupted, spring migratory activity did not reach full expression, and the gonads remained regressed (Gwinner, 1989). In such a case, a southward expansion or shift of the wintering ranges may only be possible if the photoperiodic response can be adapted to the new conditions. Other species in which photorefractoriness can be broken under conditions exceeding equatorial daylength (approx. 12.8 h), as in the garden warbler (Gwinner, 1987; Gwinner *et al.*, 1988), may succeed in extending their wintering ranges to the south.

C. Tropical Birds and Their Potential for Poleward Range Extension

For non-migratory tropical birds, a sudden shift in global climate patterns may represent a potential threat. Many tropical species living in seemingly constant environments breed seasonally and may even respond to the slight annual changes in daylength experienced close to the Equator (Hau *et al.*, 1998). How will tropical birds respond to the high-amplitude changes in daylength that are found outside current breeding ranges? Are tropical species limited in their potential for range shifts and expansions?

In African stonechats (*S. t. axillaris*), complete cycles of moult and gonadal growth can be elicited by sinusoidal photoperiodic changes (Gwinner and Scheuerlein, 1999). Equatorial stonechats behave essentially like temperate-zone stonechats (*S. t. rubecula*) when held on daylengths simulating conditions at 47.5°N latitude. Thus, the crude response to sinusoidal changes in daylength appears to be conserved in the stonechat, independent of the sub-species' origin or present migratory status. Gwinner and Scheuerlein (1999) have suggested that this potential for adaptive photoperiodic response may be a consequence of this tropical population having been founded by individuals from northern populations, or attributed to interbreeding with northern migrants from the same species. These hypotheses imply that highly adjustable photoperiodic systems can only evolve at higher latitudes and that extant tropical populations may possess photoperiodic adaptability only if they are genetically related to these populations. If this is true, many tropical bird species may be unable to alter their current distributions by adjusting to novel seasonal environments. A limited capacity for adaptive response to photoperiodic conditions outside the tropics has been experimentally demonstrated in the equatorial Andean sparrow (*Z. capensis*). Unlike its north-temperate relatives (*Z. leucophrys*, *Z. atricapilla*, and *Z. albicollis*), the Andean sparrow has no photoperiodic response mechanism for effectively preventing late summer and autumn breeding under northern seasonal conditions (Miller, 1965).

D. Birds in Urban Habitats and the Effect of Artificial Light

With increasing urbanisation and the introduction of outdoor lighting, light pollution has considerably increased in the last decades (Riegel, 1973; Cinzano *et al.*, 2001). As a result, birds living in or near highly urbanised or industrialised areas are exposed at night to increasing levels of artificial light. Several bird species have been reported to prefer habitats with night-illumination (e.g., Rees, 1982; Gorenzel and Salmon, 1995) and to be attracted, or disoriented, by artificial light sources during migration (e.g., Cochran and Graber, 1958; Drost, 1960; Avise and Crawford, 1981; Martin, 1990; Jones and Francis, 2003). Diurnal birds exposed to artificial light may become more active and extend their feeding periods into the night (e.g., Goertz *et al.*, 1980; Nein, 1989; Frey, 1993). In addition to these immediate effects, an artificial prolongation of daylength can have profound effects on the organisation of avian life cycles (Luniak and Muslow, 1988; Schmidt, 1988; Partecke, 2002). This has been most intensively studied in European blackbirds (*T. merula*). The breeding season in blackbirds living in urban habitats starts earlier and ends later than in blackbirds in rural habitats. Moreover, urban birds moult earlier and more extensively and show a reduced tendency to migrate (Luniak and Muslow, 1988; Luniak *et al.*, 1990; Partecke, 2002). Among-habitat differences in life-history traits may not only be attributed to differences in light regimes but also to differences in temperature, density, or food availability. In a common garden experiment, Partecke (2002) studied the effects of photoperiodic response and genetic differences on breeding, moult, and migration in urban and forest-dwelling blackbirds. His results suggest that among-population differences in the timing of gonadal recrudescence are predominantly an effect of different photoperiodic conditions. Differences in the timing of moult can be attributed to (genetic) differences in the response to the photoperiod. Differentiation in the amount of autumn migratory activity is most likely due to genetic differences. Thus, life-cycle adaptation to urban environments may involve phenotypic plastic responses, genetic changes in plasticity, and genetic responses.

The response of birds living in urban habitats to artificial light may have important consequences for their adaptation to climate change. Light pollution may induce earlier breeding, which may be favourable under improved conditions in spring (see above sections), and thus may facilitate adaptation to climatic amelioration. However, as anthropogenic light regimes are not necessarily related to the availability of food, the timing of breeding may be shifted towards a period in which appropriate food is scarce or missing (see, for instance, Schmidt and Steinbach, 1983; Schmidt and Einloft-Achenbach, 1984). This desynchronisation between the timing of breeding and the availability of food would lead birds into evolutionary traps (cf. Schlaepfer *et al.*, 2002), similar to the expected effect of rapid phenological change on birds living under natural photoperiodic conditions (Stenseth and Mysterud, 2002).

IV. CONCLUSION

Birds, like any other highly mobile organism, may respond to global climatic changes either by evading the altered environment or by adapting to local conditions. We stress that the response to the photoperiodic environment is a crucial factor determining to what extent range shifts and adaptive evolution will be possible. Climate change is currently altering the phenology of many organisms, but at different rates. Poikilothermic organisms may immediately respond to temperature changes. In contrast, homiothermic animals, like birds, that use temperature-independent (i.e., photoperiodic) cues to time their life cycles may get out of phase with the selective environment. In environments that have suddenly changed due to human activity, photoperiodic cues might no longer lead to adaptive responses. However, if photoperiodic conditions outside a species' or population's current range elicit phenotypic changes that are favoured by selection, range shifts and colonisation of new habitats may be facilitated. We conclude that any study on the adaptability of birds to human-caused environmental changes should take into account that adaptive processes may be facilitated or constrained by physiological and behavioural responses to the photoperiodic environment.

REFERENCES

Avise, J.C. and Crawford, R.L. (1981) *Nat. Hist.* **90**, 11–14.
Berthold, P. (1996) *Control of Bird Migration*. Chapman & Hall, London.
Berthold, P. (1998) *Naturwissenschaftl. Rundsch.* **51**, 337–346.
Berthold, P. (2001) *Bird Migration—A General Survey*. Oxford University Press.
Berthold, P., Helbig, A.J., Mohr, G. and Querner, U. (1992) *Nature* **360**, 668–670.

Blondel, J., Dias, P.C., Perret, P., Maistre, M. and Lambrechts, M.M. (1999) *Science* **285**, 1399–1402.

Both, C. and Visser, M.E. (2001) *Nature* **411**, 296–298.

Brommer, J.E. (2004) *Ann. Zool. Fennici* **41**, 391–397.

Burton, J.F. (1995) *Birds and Climate Change*. Christopher Helm, London.

Buse, A., Dury, S.J., Woodburn, R.J.M., Perrins, C.M. and Good, G.E.G. (1999) *Funct. Ecol.* **13**(**suppl. 1**), 74–82.

Cinzano, P., Falchi, F. and Elvidge, C.D. (2001) *Mon. Not. R. Astron. Soc.* **328**, 689–707.

Cochran, W.W. and Graber, R.R. (1958) *Wilson Bull.* **70**, 378–380.

Cockrem, J.F. (1995) *Reprod. Fertil. Dev.* **7**, 1–19.

Coppack, T. and Both, C. (2002) *Ardea* **90**, 369–378.

Coppack, T., Pulido, F. and Berthold, P. (2001) *Oecologia* **128**, 181–186.

Coppack, T., Pulido, F., Czisch, M., Auer, D.P. and Berthold, P. (2003) *Proc. R. Soc. Lond. B* **270**(**suppl.**), S43–S46.

Cresswell, W. and McCleary, R. (2003) *J. Anim. Ecol.* **72**, 356–366.

Crick, H.Q.P. and Sparks, T.H. (1999) *Nature* **399**, 423–424.

Daan, S., Dijkstra, C., Drent, R.H. and Meijer, T. (1988) *Proc. Int. Ornithol. Congr.* **19**, 392–407.

Dawson, A. (2002) *Ardea* **90**, 355–367.

Dawson, A., King, V.M., Bentley, G.E. and Ball, G.F. (2001) *J. Biol. Rhythms* **16**, 366–381.

Dittami, J.P. and Gwinner, E. (1985) *J. Zool. Lond.* **207**, 357–370.

Drost, R. (1960) *Proc. Int. Ornithol. Congr.* **12**, 178–192.

Engels, W.L. (1959) In: *Photoperiodism and Related Phenomena in Plants and Animals* (Ed. by R. Withrow), pp. 759–766. Publ. No. 55, American Association of Advanced Science, Washington DC.

Engels, W.L. (1961) *Biol. Bull.* **120**, 140–147.

Fiedler, W. (2003) In: *Avian Migration* (Ed. by P. Berthold, E. Gwinner and E. Sonnenschein), pp. 21–38. Springer, Berlin.

Frey, J.K. (1993) *West. Birds* **24**, 2000.

Goertz, J.W., Morris, A.S. and Morris, S.M. (1980) *Wilson Bull.* **92**, 398–399.

Gorenzel, W.P. and Salmon, T.P. (1995) *J. Wildl. Manag.* **59**, 638–645.

Gwinner, E. (1986G) *Circannual Rhythms*. Springer, Berlin.

Gwinner, E. (1987) *Ornis Scand.* **18**, 251–256.

Gwinner, E. (1989) *J. Ornithol.* **130**, 1–13.

Gwinner, E. (1996a) *Ibis* **138**, 47–63.

Gwinner, E. (1996b) *J. Exp. Biol.* **199**, 39–48.

Gwinner, E. and Scheuerlein, A. (1999) *Condor* **101**, 347–359.

Gwinner, E., Dittami, J.P. and Beldhuis, H.J.A. (1988) *J. Comp. Physiol. A* **162**, 389–396.

Hahn, T.P., Boswell, T., Wingfield, J.C. and Ball, G.F. (1997) In: *Current Ornithology, Volume 14* (Ed. by V. Nolan Jr., E.D. Ketterson and C.F. Thompson), pp. 39–80. Plenum Press, New York.

Hamner, W.M. and Stocking, J. (1970) *Ecology* **51**, 743–751.

Hau, M. (2001) *Horm. Behav.* **40**, 282–290.

Hau, M., Wikelski, M. and Wingfield, J.C. (1998) *Proc. R. Soc. Lond. B* **1391**, 89–95.

Helm, B. and Gwinner, E. (1999) *Auk* **116**, 589–603.

Helm, B. and Gwinner, E. (2001) *Avian Sci.* **1**, 31–42.

Jenni, L. and Kéry, M. (2003) *Proc. R. Soc. Lond. B* **270**, 1467–1471.

Jones, J. and Francis, C.M. (2003) *J. Avian Biol.* **34**, 328–333.

Kok, O.B., van Ee, C.A. and Nel, D.G. (1991) *Ardea* **79**, 63–65.

Lack, D. (1968) *Ecological Adaptations for Breeding in Birds.* Methuen, London.

Lambrechts, M.M., Perret, P. and Blondel, J. (1996) *Proc. R. Soc. Lond. B* **263**, 19–22.

Lambrechts, M.M., Blondel, J., Maistre, M. and Perret, P. (1997) *Proc. Natl. Acad. Sci. USA* **94**, 5153–5155.

Lambrechts, M.M., Perret, P., Maistre, M. and Blondel, J. (1999) *Proc. R. Soc. Lond. B* **266**, 1311–1315.

Langslow, D.R. (2002) In: *The Migration Atlas: Movements of the Birds of Britain and Ireland* (Ed. by C.V. Wernham, M.P. Toms, J.H. Marchant, J.A. Clark, G.M. Siriwardena and S.R. Baillie), pp. 562–564. T. & A.D. Poyser, London.

Lindström, Å., Daan, S. and Visser, G.H. (1994) *Anim. Behav.* **48**, 1173–1181.

Lofts, B. and Murton, R.K. (1968) *J. Zool. Lond.* **155**, 327–394.

Luniak, M. and Muslow, R. (1988) *Proc. Int. Ornithol. Congr.* **19**, 1787–1793.

Luniak, M., Muslow, R. and Walasz, K. (1990) In: *Urban Ecological Studies in Central and Eastern Europe* (Ed. by M. Luniak), pp. 187–200. Polish Academy of Sciences, Warsaw.

Martin, G.R. (1990) In: *Bird Migration, Physiology and Ecophysiology* (Ed. by E. Gwinner), pp. 185–197. Springer, Berlin.

McCarty, J.P. (2001) *Conserv. Biol.* **15**, 320–331.

Miller, A.H. (1960) *Proc. Int. Ornithol. Congr.* **12**, 513–522.

Miller, A.H. (1965) *Proc. Natl. Acad. Sci. USA* **54**, 97–101.

Moreau, R.E. (1972) *The Palaearctic-African Bird Migration Systems.* Academic Press, London.

Nager, R.G., Rüegger, C. and van Noordwijk, A.J. (1997) *J. Anim. Ecol.* **66**, 493–507.

Nein, R. (1989) *Beitr. Naturk. Wetterau* **9**, 213.

Nicholls, T.J., Goldsmith, A.R. and Dawson, A. (1988) *Physiol. Rev.* **68**, 133–176.

Ottosson, U., Bairlein, F. and Hjort, C. (2002) *Vogelwarte* **41**, 249–262.

Parmesan, C. and Yohe, G. (2003) *Nature* **421**, 37–42.

Partecke, J. (2002) *Annual Cycles of Urban and Forest-living European Blackbirds* (Turdus merula): *Genetic Differences or Phenotypic Plasticity?* PhD thesis, University of Munich.

Perrins, C.M. (1970) *Ibis* **112**, 242–255.

Pigliucci, M. (2001) *Phenotypic plasticity. Beyond Nature and Nurture.* Johns Hopkins University Press, Baltimore, MD.

Pohl, H. (1999) *Physiol. Behav.* **67**, 327–337.

Pulido, F. (2000) *Evolutionary Quantitative Genetics of Migratory Restlessness in the Blackcap (Sylvia atricapilla).* Tectum, Marburg.

Pulido, F. and Berthold, P. (2003) In: *Avian Migration* (Ed. by P. Berthold, E. Gwinner and E. Sonnenschein), pp. 53–77. Springer, Berlin.

Pulido, F. and Coppack, T. (2004) *Anim. Behav.* **68**, 167–173.

Pulido, F., Coppack, T. and Berthold, P. (2001a) *Ring* **23**, 149–157.

Pulido, F., Berthold, P., Mohr, G. and Querner, U. (2001b) *Proc. R. Soc. Lond. B* **268**, 953–959.

Rees, E.C. (1982) *Wildfowl* **33**, 119–132.

Rees, E.C. (1989) *Anim. Behav.* **38**, 384–393.

Riegel, K.W. (1973) *Science* **179**, 1285–1291.

Root, T.L. (1988a) *J. Biogeogr.* **15**, 489–505.

Root, T.L. (1988b) *Ecology* **69**, 330–339.

Root, T.L. (1993) In: *Biotic Interactions and Global Change* (Ed. by P.M. Kareiva, J.G. Kingsolver and R.B. Huey), pp. 280–292. Sinauer, Sunderland, Massachusetts.

Root, T.L., Price, J.T., Hall, K.R., Schneider, S.H., Rosenzweig, C. and Pounds, J.A. (2003) *Nature* **421**, 57–60.

Rowan, W. (1925) *Nature* **115**, 494–495.

Sanz, J.J., Potti, J., Moreno, J., Merino, S. and Frías, O. (2003) *Glob. Change Biol.* **9**, 461–472.

Schlaepfer, M.A., Runge, M.C. and Sherman, P.W. (2002) *Trends Ecol. Evol.* **17**, 474–480.

Schmidt, K.-H. (1988) *Proc. Int. Ornithol. Congr.* **19**, 1795–1801.

Schmidt, K.-H. and Einloft-Achenbach, H. (1984) *Vogelwelt* **105**, 97–105.

Schmidt, K.-H. and Steinbach, J. (1983) *J. Ornithol.* **124**, 81–83.

Silverin, B. (1994) *Ethol. Ecol. Evol.* **6**, 131–157.

Silverin, B., Massa, R. and Stokkan, K.A. (1993) *Gen. Comp. Endocrinol.* **90**, 14–22.

Sparks, T.H. and Menzel, A. (2002) *Int. J. Climatol.* **22**, 1715–1725.

Stenseth, N.C. and Mysterud, A. (2002) *Proc. Natl. Acad. Sci.* **99**, 13379–13381.

Styrsky, J.D., Berthold, P. and Robinson, D. (2004) *Anim. Behav.* **67**, 1141–1149.

Terrill, S.B. and Berthold, P. (1990) *Oecologia* **85**, 266–270.

Thomas, C.D. and Lennon, J.J. (1999) *Nature* **399**, 213.

Thomas, D.W., Blondel, J., Perret, P., Lambrechts, M.M. and Speakman, J.R. (2001) *Science* **291**, 2598–2600.

Valiela, I. and Bowen, J.L. (2003) *AMBIO* **32**, 476–480.

van Noordwijk, A.J., McCleery, R.H. and Perrins, C.M. (1995) *J. Anim. Ecol.* **64**, 451–458.

Visser, M.E. and Holleman, L.J.M. (2001) *Proc. R. Soc. Lond. B* **268**, 289–294.

Visser, M.E. and Lambrechts, M.M. (1999) *Proc. Int. Ornithol. Congr.* **22**, 249–264.

Visser, M.E., van Noordwijk, A.J., Tinbergen, J.M. and Lessells, C.M. (1998) *Proc. R. Soc. Lond. B* **265**, 1867–1870.

Walther, G.-R., Post, E., Convey, P., Menzel, A., Parmesan, C., Beebee, T.J.C., Fromentin, J.-M., Høgh-Guldberg, O. and Bairlein, F. (2002) *Nature* **416**, 389–395.

Widmer, M. (1999) *Altitudinal Variation of Migratory Traits in the Garden Warbler Sylvia borin*. PhD thesis, University of Zürich.

Williams, T.D., Dawson, A., Nicholls, T.J. and Goldsmith, A.R. (1987) *J. Reprod. Fert.* **80**, 327–333.

Wingfield, J.C. (1980) In: *Avian Endocrinology* (Ed. by A. Epple and M.H. Stetson), pp. 367–389. Academic Press, New York.

Wingfield, J.C., Hahn, T.P., Levin, R. and Honey, P. (1992) *J. Exp. Zool.* **261**, 214–231.

Wingfield, J.C., Doak, D. and Hahn, T.P. (1993) In: *Avian Endocrinology* (Ed. by J. Sharp), pp. 111–122. Soc. Endocrinol., Bristol.

Microevolutionary Response to Climatic Change

FRANCISCO PULIDO* AND PETER BERTHOLD

I. SUMMARY

Organisms may respond to changing environments by evading the new conditions or by adapting to them. Recently, a large body of evidence has been collected indicating that phenotypic adaptation to climate change is widespread. Adaptation may be achieved by phenotypic adjustment or by changes in the genetic composition of populations. Both processes can assure the survival of populations in changing environments, but at different time scales and at different costs. Recent studies indicate that the mechanisms leading to

---

E-mail address: pulido@orn.mpg.de (F. Pulido)

ADVANCES IN ECOLOGICAL RESEARCH, VOL. 35
0065-2504/04 $35.00 DOI 10.1016/S0065-2504(04)35008-7

adaptive phenotypic changes in birds may be complex, involving both plastic response and genetic change. Changes in the timing of breeding, for instance, seem to be predominantly caused by phenotypic adjustment to environmental conditions. Shifts in the genetic composition of populations have been demonstrated to be involved in recent changes in morphology and migratory behaviour. The presence of considerable amounts of additive genetic variation within and among avian populations, and examples of rapid evolutionary response to rare climatic events suggest that birds have a high potential for adaptive evolutionary change. However, it is presently unclear whether this is a general pattern, and which factors actually limit the adaptability of avian populations. Antagonistic genetic correlations and maladaptive phenotypic responses (evolutionary traps) are probably the most important constrains to microevolutionary change. Furthermore, the loss of genetic variation due to population declines, and gene flow in the presence of among-population variation in the response to climate change may limit the rate of adaptive evolution. Future research should try to identify the targets of selection and gauge the importance of constraints to microevolutionary change.

II. INTRODUCTION

Currently, climate is changing at an unprecedented rate. Present-day climatic change is characterised by a rapid global increase in temperature, whereby temperatures are not increasing evenly throughout the year and not at the same rate and extent geographically. In the northern hemisphere, temperature increase is higher in winter and early spring than in summer or autumn, and considerably larger at high altitudes and latitudes than at low altitudes and in the south (Houghton et al., 2001). But locally these trends may not be apparent or may even be reversed (e.g., Kozlov and Berlina, 2002). The global rise in temperature has caused phenological changes (Menzel and Fabian, 1999; Menzel, 2000; Sparks and Menzel, 2002), shifts in climate zones (Fraedrich et al., 2001), and changes in the frequency of extreme weather events like droughts, periods of extreme heat, storms and floods (Easterling et al., 2000a,b; Meehl et al., 2000).

These environmental changes impose major challenges to animal and plant populations, and may potentially lead to their extinction. Organisms may respond to this climatic change in different ways (Holt, 1990). One possible response is to evade the unfavourable environment. This may result in shifts or expansions of ranges, which has been documented for various plant and animal species (Parmesan, 1996; Parmesan et al., 1999), including birds (e.g., Burton, 1995; Thomas and Lennon, 1999; Valiela and Bowen, 2003). Highly mobile organisms, like birds, are expected to more readily respond to environmental change by this means than sessile species and species with low potential of dispersal. However, the availability of appropriate food at the newly colonised site may limit this process (Root, 1993; Huntley, 1994; Price and Glick, 2001). Irrespective of whether organisms colonise a new area, or if they stay within their traditional ranges, populations have to respond to environmental change by adapting to the new environmental conditions (Geber and Dawson, 1993). Traditionally, adaptation is considered to be the result of genetic change in response to selection (see, for instance, Futuyma, 1998), i.e., microevolution. This process of adaptive evolution is relatively slow (but see Hendry and Kinnison, 1999; Rice and Emery, 2003), and may prevent extinction only if selection is not too strong and the lag between the population mean and the optimal phenotype does not become too large (Lynch and Lande, 1993; Lande and Shannon, 1996). Alternatively, or in addition, adaptation to new environmental conditions can be achieved by phenotypic changes of the individual, i.e., phenotypic plasticity (Meyers and Bull, 2002). Individual adjustment may be reversible or irreversible and may or may not be genetically fixed (Piersma and Drent, 2003), in which case it may respond to selection (Pigliucci, 2001; Scheiner, 2002). The evolution of phenotypic plasticity, or its evolutionary change, may be constrained by the costs of plasticity (DeWitt *et al.*, 1998; Pigliucci, 2001) and the restrictive selection conditions needed (Scheiner and Lyman, 1991; Scheiner, 1993, 2002). Moreover, moderate levels of plasticity will facilitate evolutionary change (Price *et al.*, 2003).

Whether a population will respond to climatic change by phenotypic plasticity or by genetic change seems, at first sight, equally adequate. However, there are some fundamental differences in the consequences of these modes of adaptation. Adaptive responses by phenotypic plasticity are limited by the range of environments in which phenotypic responses are adaptive. Beyond this range phenotypic adjustment may be insufficient or maladaptive (see Coppack and Pulido, 2004, this volume). In a continuously changing environment, adaptation by phenotypic plasticity may, thus, sooner or later lag behind environmental changes, which will confer fitness consequences for the population. Microevolutionary change in response to natural selection can, in principle, enable populations to adapt to ever changing environments. The long-term response to selection is only limited by the amount of additive genetic variation present in the population (Bradshaw, 1991; Bradshaw and McNeilly, 1991). Even this limitation may not be a serious problem as new genetic variation can arise by mutation and transformation of epistatic and dominance variance into additive genetic variation (see Roff, 1997), as indicated by the persisting response in long-term artificial selection experiments (see, for instance, Yoo, 1980 and review by Hill and Caballero, 1992). However, the rate of adaptive evolution critically depends on generation time, and may be very slow in long-lived organisms with overlapping generations (A.J. van Noordwijk, pers. comm.). Therefore, organisms with long generation times and a low potential for population growth may not be able to adapt to rapidly changing environments (see Reznick and Ghalambor, 2001).

Climatic changes, and associated alterations in habitats and biotic interactions have probably been the most important causes for natural selection in the wild, and for this reason one of the driving forces in evolution (Pearson, 1978; Endler, 1986). In a recent review on the biological consequences of global warming, Hughes (2000) predicted changes by four different mechanisms: direct physiological effects, effects on the distribution of animals, phenological changes and adaptive evolution. While he gave a number of examples for all categories of changes, he did not even treat evolutionary changes separately, and gave only one example (Rodríguez-Trelles and Rodríguez, 1998). Although from an evolutionary perspective this categorisation is not intelligible, as physiological and phenological effects may well be a consequence of evolutionary change, it is clear that convincing instances for evolutionary change in response to extant climatic shifts are very rare, and at that time were virtually lacking. Since that review, two studies have been published that have demonstrated that natural populations do respond to climatic warming by evolutionary change (Bradshaw and Holzapfel, 2001; Réale et al., 2003). In addition, a number of studies have investigated the causes underlying phenotypic changes in natural populations and the mechanisms responsible for the lack of evolutionary response (e.g., Etterson and Shaw, 2001; Kruuk et al., 2001; Merilä et al., 2001a,b; Sheldon et al., 2003). This work and a number of other studies that are currently under way constitute the fundaments for understanding adaptation to climate change. Although we are only at the beginning of this research, the growing number of studies recently published on this subject justifies writing a review on the evolutionary response to global warming based on empirical results. Because birds are the group of organisms for which we have the best long-term data sets on the composition and behaviour of wild populations, they are central to our understanding of the mechanisms of evolutionary change. In contrast to previous reviews on the evolutionary response to climate change (Holt, 1990; Bradshaw and McNeilly, 1991; Geber and Dawson, 1993; Hoffmann and Blows, 1993; Travis and Futuyma, 1993; Rodríguez-Trelles et al., 1998) that have emphasised the potential and limits of adaptive response, the aim of this chapter is to review recent empirical work that has investigated microevolutionary response to climate change and to discuss potentials and shortcomings of different approaches. We shall, in particular, focus on the evolutionary response of avian migratory behaviour.

III. MICROEVOLUTIONARY CHANGE—MECHANISMS AND APPROACHES

Current climatic changes have had, and are persistently having, a profound impact on animal and plant populations (McCarty, 2001; Walther *et al.*, 2002; Parmesan and Yohe, 2003; Root *et al.*, 2003). The most prominent changes in birds have been changes in the timing of breeding (e.g., Brown *et al.*, 1999; Crick *et al.*, 1997; Dunn, 2004, this volume) and migration (e.g., Tryjanowski *et al.*, 2002; Butler, 2003; Hüppop and Hüppop, 2003; Jenni and Kéry, 2003; Lehikoinen *et al.*, 2004, this volume; Fiedler *et al.*, 2004, this volume), in clutch size and the number of clutches (e.g., Winkel and Hudde, 1996, 1997; Møller, 2002), in migration distance and the propensity to migrate (reviewed by Berthold, 1998a; Fiedler, 2003), and in body size (e.g., Ludwichowski, 1997; Jakober and Stauber, 2000; Yom-Tov, 2001). These changes in bird populations have been interpreted as being adaptive responses to climate change. However, most of these studies have measured phenotypic change alone without separating genetic and environmental components contributing to that change (Hendry and Kinnison, 1999). Similarly, palaeontologists have assumed that phenotypic changes represent evolutionary change, as they cannot separate the causes of differentiation (Travis and Futuyma, 1993; Barnosky *et al.*, 2003).

But to what extent is this assertion correct, or is phenotypic change primarily a consequence of phenotypic plasticity? A recent meta-analysis of rates of change at the genotypic and phenotypic level revealed that the degree of diversification was lower in phenotypic than in genetic studies. One explanation for this finding is that "phenotypic plasticity may make an important contribution to the earliest stages of population divergence or evolution" (Kinnison and Hendry, 2001). Diversification by genetic change may be initially slower than by phenotypic plasticity, but then proceed at a more constant rate over a longer time interval, therefore, on the long run, the net rate of change is expected to be higher (Trussel and Etter, 2001; Pulido, 2004). Furthermore, phenotypic plasticity has been shown to be ubiquitous and to play an important role in the evolutionary response to environmental change (Pigliucci, 2001; Price *et al.*, 2003; West-Eberhard, 2003; Coppack and Pulido, 2004, this volume).

Evolutionary change is by definition the "change over time of the proportion of individual organisms differing genetically in one or more traits" (Futuyma, 1998). Thus, to demonstrate that phenotypic changes are a result of evolutionary change, we have to demonstrate that phenotypic differences have a genetic basis. A number of different methods have been proposed and applied for ascertaining genetic differentiation in space and time (see Reznick and Travis, 1996; Reznick and Ghalambor, 2001; Boake *et al.*, 2002; Conner, 2003). Among the most commonly used methods are common-garden experiments, reciprocal transplants, reciprocal crossings, artificial selection experiments, longitudinal studies allowing to assess individual phenotypic change, and the estimation of family resemblances from which different quantitative genetic parameters can be derived (e.g., heritability, additive genetic variance, individual estimated breeding values). Moreover, evolutionary change may be demonstrated by showing that actual phenotypic change is in accord with changes expected from selection intensities, genetic variation and among-trait covariation within populations (Grant and Grant, 1995, 2002; Roff and Fairbairn, 1999). None of these methods is without problems, as common-environment effects, including maternal effects, and genotype-by-environment interactions may confound results and suggest genetic differentiation where there is none (Møller and Merilä, 2004, this volume).

Two of these methods have become particularly useful in the study of evolutionary change in response to climate change: common-garden experiments over time and the calculation of estimated breeding values.

Common-garden experiments are based on the idea that if individuals that have been sampled in different areas or at different times express phenotypic differences when held under identical, controlled environmental conditions, these differences are likely to have a genetic basis. The accuracy of this method very much depends on whether it is possible to keep the environment constant and to exactly replicate the conditions of measurement over a long period of time. A group of individuals, which is known not to have changed genetically over time, may serve as a control, and help to minimise this error. Another potentially important shortcoming of this method is its sensitivity to maternal effects. If the phenotype of an individual is not only determined by its genotype and by the environmental conditions it has experienced but also by the phenotype of its mother (including environmental effects), we talk of maternal effects to the phenotype. These effects can be long-lasting and may not be eroded even when individuals are transferred to a common environment very early in life (see reviews in Mousseau and Fox, 1998). This problem may be circumvented if common-garden experiments are combined with breeding experiments where phenotypic differences in the common environment are assessed in the generation of individuals produced by mothers held in the controlled environment, i.e., in the F_1 generation. Strong and persistent maternal effects, however, may not be totally removed after one "round of breeding". The breeding of an F_2 generation in a common environment will in most cases be sufficient to minimise these effects. Because of the difficulties associated with this method it is particularly well suited for monitoring organisms for which controls can easily be preserved (e.g., clonal organism), and for studying traits which are known to be influenced only by few environmental variables—preferably only one—that can be experimentally controlled.

A common garden approach over time was used by Bradshaw and Holzapfel (2001) to demonstrate microevolutionary response of photoperiodism in pitcher plant mosquitoes *Wyeomyia smithii*. They collected mosquitoes from 31 locations covering more than 20 degrees of latitude in 4 years from 1972 through 1996, and measured in two different approaches critical daylengths of pupation and of the initiation and maintenance of diapause under controlled conditions. The comparisons of the changes between 1972 and 1996 and between 1988 and 1993 yielded similar results. In both periods the critical photoperiod had changed towards shorter daylengths, whereby changes in the northern populations were more pronounced than in southern populations. These results were in accord with temperature changes in North America that have been strongest at highest latitudes. The observed change in the critical photoperiod between 1972 and 1999 was equivalent to a delay of diapause by 9 days. Although the data had been collected for other reasons than for studying evolutionary change, and controls were not kept, this study is a convincing example for the use of a controlled environment to elucidate genetic variation in time and space. We have used a similar approach to study evolutionary change in migratory behaviour in a bird population (see below).

An alternative approach for studying the causes of phenotypic change over time is the estimation of breeding values of individuals. The breeding value is a statistical measure of the "individual genotype" in a particular population, and is defined as "the sum of the average effects of the genes an individual carries for a trait" (Falconer and Mackay, 1996). Breeding values have previously been estimated by animal and plant breeders for identifying those individuals with the prospect of yielding the highest selection gain. With the development of new statistical techniques (Henderson, 1986; Knott *et al.*, 1995) and the increase of computer power it is now possible to estimate breeding values for large, unbalanced data sets, like those obtained in long-term studies on natural populations in the field (Kruuk, 2004). One of the advantages of these new statistical approaches is that different genetic (e.g., maternal, paternal) and environmental effects (condition, age, food availability, temperature, etc.) can be estimated separately, and their importance can be evaluated. The accuracy of breeding value estimates critically depends on the accuracy of the quantitative genetic estimates (i.e., genetic variances and covariances). It is, therefore, potentially dependent on family size, pedigree size and complexity, and the number and diversity of environments considered. However, to our knowledge, biases and limitation of this approach for detecting and quantifying microevolutionary change in the wild have hitherto not been studied, and have remained largely unconsidered in the interpretation of results.

The study of the correlation between breeding values and fitness has been shown to predict selection intensities and responses more accurately than the classical approach based on the correlation between phenotypic values and fitness because selection gradients based on breeding values are not biased by environmental covariation (Rausher, 1992; Kruuk *et al.*, 2003). For this reason, this method has provided new insight into the process of natural selection in wild populations by detecting cryptic evolutionary change (Merilä *et al.*, 2001a), and the cause for the lack of evolutionary change in the presence of additive genetic variation and selection (Kruuk *et al.*, 2002). It has recently become the most important tool in the study of evolutionary response to climate change. This method has been applied, for instance, to investigate changes in antler size in red deer *Cervus elaphus* (Kruuk *et al.*, 2002), parturition date in red squirrels *Tamiasciurus hudsonicus* (Réale *et al.*, 2003), tarsus length (Kruuk *et al.*, 2001), body weight at fledging (Merilä *et al.*, 2001a,b), clutch size and timing of breeding in the collared flycatcher *Ficedula albicollis* (Sheldon *et al.*, 2003), and timing of spring arrival in barn swallows *Hirundo rustica* (A.P. Møller, unpubl.; see below).

Merilä *et al.* (2001b) investigated the change in condition in the collared flycatcher population of Gotland (Sweden) from 1980 through 1999. They found a phenotypic decrease in condition over time despite selection favouring individuals in better condition and the presence of significant amounts of additive genetic variation for this trait (Figure 1a). However, mean estimated breeding values for condition increased over time as expected from the direction and strength of selection (Figure 1b). This discrepancy between changes in condition at the phenotypic and at the genotypic level can be explained by changes in the environment that may induce changes in opposite direction to evolutionary change (countergradient variation). In the case of the Gotland flycatcher population, the deterioration of environmental condition was probably due to a reduction of the availability of caterpillars—the primary food of nestlings—as a consequence of an increasing asynchrony between the caterpillars and bud burst of their host trees, i.e., oaks (Visser and Holleman, 2001).

It is currently unknown how widespread this phenomenon is. There are a number of studies that have reported countergradient variation (Conover and Schultz, 1995), but because this kind of work needs long-term data sets on individually marked related individuals, there are only few studies that have been able to investigate this phenomenon. Irrespective of whether this phenomenon is common or not, this study cautions against drawing conclusions about evolutionary change based on phenotypic change alone. In the presence of countergradient variation, phenotypic change in one direction may be paralleled by strong evolutionary changes in that direction, no evolutionary change, or genetic changes in the opposite direction. Without knowing how environmental variation influences the expression of a trait and whether there have been changes in the relevant environmental variables, no inferences on evolutionary response may be drawn from phenotypic changes.

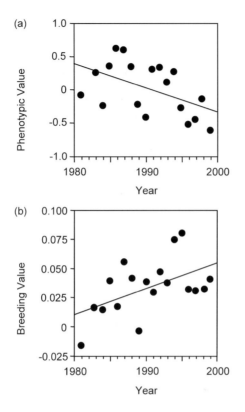

Figure 1 Observed annual changes in body condition in a collared flycatcher population from Gotland (Sweden). (a) Changes in mean phenotypic value of condition index. (b) Changes in mean estimated breeding values of condition index (from Merilä *et al.*, 2001c, with permission from *Nature*).

IV. THE PROSPECTIVE IMPORTANCE OF DIFFERENT EVOLUTIONARY PROCESSES TO ADAPTIVE EVOLUTIONARY CHANGE

A. The Response of Bird Populations to Natural Selection

There is probably no other group of organisms in which the impact of natural selection on trait distribution in natural populations has been studied in more detail than in birds. One reason for this is that the offspring of birds can be easily marked individually and trait distributions and fitness, as measured, for instance, as lifetime reproductive success, can be followed over several generations. Selection studies in natural bird populations have shown that extreme weather events are probably the most important and strongest selective agents in nature (Price and Boag, 1987).

One of the first, and probably the best analysed, studies of natural selection in the wild was the study by Bumpus (1899) who investigated the effects of a snow storm on morphological trait distribution in house sparrows *Passer domesticus*, overwintering in Rhode Island. He measured 10 morphological traits in the sample of birds that had died and in birds that had survived the storm. He found that birds of intermediate size were those that had survived best. Later analyses revealed among-sex differences in selection, e.g., directional selection for large males (e.g., Johnston *et al.*, 1972; Lande and Arnold, 1983). This approach of measuring trait distributions before and after a selection event has been the prevailing method used to detect and reconstruct selection. If, in addition, genetic variances and covariances for the traits under scrutiny are known, selection intensities can be estimated (Lande and Arnold, 1983; Arnold, 1988, 1994; for a review of selection intensities in the wild, see Kingsolver *et al.*, 2001). However, one major problem with this method is that the traits under selection may not be measured. Moreover, the samples taken before and after the selection event may not be representative, and immigration after the selection event may bias trait values in the presumed "survivors" (Price and Yeh, 1999; Price *et al.*, 2000; Møller and Merilä, 2004, this volume). Another problem is that selection may act on environmental deviations, or that differential survival may be a consequence of environmental covariation between trait expression and fitness (see above; Merilä *et al.*, 2001c; Kruuk *et al.*, 2003).

Using this approach a number of selection events caused by extreme climatic events have been studied in wild bird populations (see references in Price and Boag, 1987), of which the best known are the studies by Peter and Rosemary Grant on the medium ground finch *Geospizia fortis* and the cactus finch *G. scandens* on the Galapagos island Daphne Major (Grant and Grant, 1989, 1993, 2002). In a nutshell, the results of these studies can be summarised as follows: (1) Evolutionary response to extreme climatic events can be considerable and (2) high evolutionary rates are both due to high selection intensities and high heritabilities. Moreover, studies following more than one selection episode indicate that selection often fluctuates in direction and intensity and, therefore, over longer periods of time, rates of evolutionary change are much lower than predicted from one selection episode (e.g., Grant and Grant, 2002; see also meta-analysis by Kinnison and Hendry, 2001).

Although many studies suggest that selection imposed by climatic change may lead to very rapid evolutionary changes, it is doubtful whether these results can be generalised. First, most studies have investigated morphological traits which are known to have high heritabilities (e.g., Mousseau and Roff, 1987;

Roff, 1997). Second, selection has been predominantly studied after extreme climatic events (cold spells, droughts, flooding, etc.). The response to such strong selection, however, may differ substantially from continuous directional selection. While extreme events often involve high mortality of birds and predominantly affect viability selection, persistent selection is often weaker and may involve fitness differences in both viability and fecundity (e.g., Sheldon *et al.*, 2003). Moreover, under less extreme environmental changes, adjustment at the individual level (e.g., by phenotypic plasticity) is likely to be more important (Meyers and Bull, 2002).

B. The Effects of Gene Flow

As a consequence of habitat changes, range expansions and shifts, we expect patterns of gene flow to change. Populations previously separated by distance, by ecological or geographic barriers may come into contact and interbreed (see Rhymer and Simberloff, 1996). The strength of the effects of such a mixing of populations will depend on the degree of introgression (i.e., mating system), among-population differentiation, effective population size and mutation rates (Edmands and Timmerman, 2003). One possible consequence of population mixing could be a reduction in fitness as a consequence of the disruption of local co-adapted gene complexes and the occurrence of detrimental gene interactions, i.e., outbreeding depression (Frankham *et al.*, 2002). Moreover, gene flow may constrain local adaptation (Slatkin, 1987; Storfer, 1999; Lenormand, 2002). However, gene flow may also increase fitness by increase genetic variation and disrupting unfavourable genetic correlations (Grant and Grant, 1994). This could facilitate rapid evolution. It is currently not known whether these effects of gene flow will be important in the adaptation to climate change, or as a consequence of changes in breeding ranges.

Yet, gene flow may become an important component for the adaptation to climate change by another mechanism. The expected directional expansion of ranges of populations dwelling at lower latitudes to higher latitudes, and from lower altitudes to higher altitudes, will most probably cause asymmetrical gene flow from populations adapted to warmer climatic conditions to populations adapted to colder environments. We predict that this inflow of pre-adapted genotypes will facilitate evolutionary adaptation to novel environmental conditions. Although we currently have no evidence for such adaptive unidirectional gene flow it may not be uncommon, but often overlooked, as evolutionary changes in response to local environmental change will have the same effect. Furthermore, gene flow could be reduced as a consequence of a reduction of migratoriness in northern populations (see below; Møller and Merilä, 2004, this volume), as dispersal and geographic differentiation are associated with migratoriness in birds (see Paradis *et al.*, 1998; Belliure *et al.*, 2000). This process could accelerate local adaptation by reducing introgression of genes from migratory to other populations.

Global climate change is also likely to increase the frequency of interspecific hybridisation. As a consequence of differential effects on population sizes in different species, abundances of some species will dramatically increase while other species will decline to the brink of extinction (Berthold, 1990; McCarty, 2001; McLaughlin *et al.*, 2002; Lemoine and Böhning-Gaese, 2003). Under the situation that one of two potentially hybridising species is rare and the other common, the frequency of hybridisation is expected to increase (Wirtz, 1999; Randler, 2002). Hybridisation can have dramatic effects, particularly on the rarer species, extending its evolutionary and ecological potential. This has been extensively studied in Galapagos finches. Here, climatic change has been shown to increase the rate of hybridisation by changing the abundance of intermediate seed sizes, which can best be exploited by hybrids (Grant and Grant, 1992, 1994, 1996). In other species there is also evidence for introgressive hybridisation as a consequence of changes in population sizes, for instance, in European redstarts (*Phoenicurus phoenicurus* and *P. ochruros*; Berthold *et al.*, 1996; Grosch, 2003), and warblers (*Hippolais icterina* and *H. polyglotta*; Faivre *et al.*, 1999). These instances need, however, to be verified by molecular methods, and may not hold under closer scrutiny (S. Bensch, pers. comm.).

C. The Effects of Inbreeding

If populations suffer rapid and strong declines and when, thereafter, population sizes remain small for a longer period, or when new populations are founded by a few individuals, inbreeding coefficients are likely to increase. Elevated levels of inbreeding may have negative, fitness-reducing effects ultimately leading to the extinction of populations (Frankham, 1995a; Keller and Waller, 2002). In birds, inbreeding is associated with reduced hatching success, reduced survival and lower recruitment into the next generation (e.g., Kempenaers *et al.*, 1996; Daniels and Walters, 2000; Keller *et al.*, 2002; Spottiswoode and Møller, 2004). Even moderate population bottlenecks (<600 individuals) can have significant fitness consequences for many bird populations (Briskie and Mackintosh, 2004). However, population declines will not inevitably cause an increase of the level of inbreeding if the more inbred individuals do not survive population crashes. This has been shown in the song sparrow *Melospiza melodia* population on Mandarte Island (Keller *et al.*, 1994). If population crashes increase as a consequence of increasing frequency of extreme climatic events, recessive deleterious alleles causing inbreeding depression could be purged from the population and reduce inbreeding effects (e.g., Templeton and Read, 1984; Barrett and Charlesworth, 1991). But purging may be efficient only under particular environmental conditions, and may not work under changing or stressful environmental conditions (e.g., Bijlsma *et al.*, 2000; Kirstensen *et al.*, 2003). Thus, inbreeding effects due to population bottlenecks may persist for long periods of time, as was shown in birds introduced to New Zealand (Briskie and Mackintosh, 2004), probably because they are caused by a reduction of "overdominance effects" rather than by the expression of recessive deleterious alleles.

Another effect of inbreeding has been recently found in an endangered North American breeding bird: the red-cockaded woodpecker *Picoides borealis* (Schiegg *et al.*, 2002). During the last decades, laying date has significantly advanced in two populations of this species. This was most probably due to individual adjustment of the timing of breeding to changes in temperatures and the timing of food availability. Breeding experience, the age of the mother and the level of inbreeding affected the adjustment of laying date. A trend for earlier laying was found among non-inbred birds, but not among inbred individuals, suggesting that the effect of climate on laying date depends on the level of inbreeding. It is unclear why inbred female red-cockaded woodpeckers do not adjust to climate change. One possible explanation would be that inbreeding reduces plasticity, because inbred individuals are in inferior condition and cannot allocate energy into plasticity. Low immunocompetence, which has been found to be associated with inbreeding in birds (Reid *et al.*, 2003), may affect the timing of breeding, as recently shown in tree swallows *Tachycineta bicolor* (Hasselquist *et al.*, 2001). Unfortunately, experiments testing for the association between levels of inbreeding and phenotypic plasticity are lacking.

V. CHANGES IN THE LEVEL OF GENETIC VARIATION

It is not clear how climate change will affect the level of genetic variation in natural populations. To what extent changes are to be expected depends on a number of factors, primarily on the form, strength and constancy of selection, on the size of the population and population trends, and on the degree of isolation, i.e., gene flow (Lacy, 1987; Booy *et al.*, 2000). Furthermore, sex ratio and mating system are major determinants of effective populations size (e.g., Frankham, 1995b; Nunney, 1995). Although we currently have some evidence that sex ratio in birds can be adaptively modified (but see Ewen *et al.*, 2004, this volume), we do not know if and in which direction changes could take place. Potential factors influencing the sex ratio are the quality of territories (i.e., resource availability), and the quality of males (Sheldon, 1998). Mating system is also highly dependent on resource availability. The higher the availability of resources, the higher is the level of promiscuity in the population (Forstmeier et al., 2001; Leisler et al., 2002). A reduction of food availability or quality may thus result in lower effective population size not only because of the reduction of the carrying capacity of an area but also because of changes in mating systems. Habitat deterioration in combination with habitat loss, and the increase of fragmentation, will increase the effects of inbreeding and genetic drift by reducing population size and decreasing gene flow (e.g., Gibbs, 2001). Moreover, persistent natural selection as imposed by ongoing climate changes is accompanied by a reduction of population fitness, because adaptive evolution will lag behind rapid environmental changes, and the optimal phenotype cannot evolve in time (Lynch and Lande, 1993; Gomulkiewicz and Holt, 1995; Nunney, 2003). The increase of the frequency of rare climatic events will exert strong viability selection that may drastically reduce population sizes (to the point of extinction). As a consequence, the likelihood for the loss of genetic variation by genetic drift and inbreeding will increase. The potential for adaptive response to climate change may be further reduced by anthropogenic changes of quality, size and distribution of suitable habitats. Many bird populations have already suffered declines as a consequence of habitat deterioration and destruction (Bauer and Berthold, 1997). Population declines and habitat fragmentation are consequences of this process that will reduce the genetic variability within populations and thereby their evolvability and fitness (Booy et al., 2000; Gaggiotti, 2003; Reed and Frankham, 2003). Generally, we predict that populations in disturbed and fragmented habitats are less likely to survive environmental changes as both their potential for adapting to new conditions *in situ* and the potential for evading these conditions by dispersal to areas with more favourable environmental conditions, i.e., range shifts, are restricted (Simberloff, 1995; Travis, 2003). Moreover, competition with invasive species may further accelerate population declines, especially in endemic species and ecological specialists (Benning et al., 2002; McLaughlin et al., 2002). It is controversial whether peripheral and isolated population will be able to endure new climatic conditions because they are adapted to resist extreme climatic conditions and environmental perturbation (Safriel et al., 1994; Reed et al., 2003), or if, alternatively, they will be the first to go extinct because of increased levels of stress and reduced levels of genetic variation (e.g., Parsons, 1990; Hoffmann and Parsons, 1997; Hoffmann et al., 2003). Probably one crucial determinant of the evolutionary potential of peripheral populations is the time they had for evolving genetic adaptations to transition areas and sub-optimal habitats, and whether they could maintain a sufficiently large effective population size over evolutionary time (Lesica and Allendorf, 1995).

VI. ADAPTIVE CHANGES IN LAYING DATE IN RESPONSE TO CLIMATE CHANGE

Probably, the best studied and supported effect of climate change on birds is the trend for earlier egg laying (Winkel and Hudde, 1996, 1997; Crick *et al.*, 1997; Brown *et al.*, 1999; Crick and Sparks, 1999; Dunn and Winkler, 1999; Koike and Higuchi, 2002; Sergio, 2003; Visser *et al.*, 2003). This trait is known to have a significant heritability in many populations (see Boag and van Noordwijk, 1987), and to respond to selection (Flux and Flux, 1982). But laying date seems also to be a trait which is adjusted individually to local conditions, both by phenotypic plasticity and by individual learning (e.g., van Noordwijk and Müller, 1994; Juillard *et al.*, 1997; Grieco *et al.*, 2002).

All studies that have investigated the causes of changes in laying date have found that currently observed trends for earlier laying are best explained by individual adjustment to increasing temperatures (Przybylo *et al.*, 2000; Both and Visser, 2001; Schiegg *et al.*, 2002; Sergio, 2003; Sheldon *et al.*, 2003). Convincing evidence for this conclusion has been provided by longitudinal population studies (Przybylo *et al.*, 2000; Both and Visser, 2001). They show that the response of individual birds to among-year variation in temperature is not different from the populational response in laying date. As a consequence, no change in the genetic composition of the population needs to be inferred to explain the trend for earlier laying. The most detailed study to date on the evolution and causes of phenotypic change in response to climatic variation is the study by Sheldon *et al.* (2003) that investigated phenotypic and genetic changes in the timing of breeding and clutch size from 1988 to 1999 in the collared flycatcher population from Gotland (Sweden). Despite using large sample sizes and powerful statistical techniques, no evidence for evolutionary change towards earlier laying was found in this study. The previous finding that among-year correlation between NAO and laying date is caused by phenotypic plasticity (Przybylo *et al.*, 2000) was confirmed.

These results on adaptive changes in the timing of breeding in birds are very consistent among studies and seem to indicate that there are general principles in the mode of adaptation of temperate-zone birds to climate change. We currently do not know whether this high phenotypic flexibility in the timing of breeding is also found in birds breeding in habitats with less among-year variation of weather conditions (e.g., in the tropics). Moreover, there seem to be limits to the adjustment of the timing of breeding to climatic conditions, as has recently been shown in a migratory bird species (Both and Visser, 2001). Constraints on the adjustment of laying date may be determined by the flexibility of the phase relationship between timing of spring migration, breeding and post-juvenile and post-nuptial moult (Coppack and Pulido, 2004, this volume; Pulido and Coppack, 2004).

VII. MICROEVOLUTIONARY RESPONSE OF MIGRATORY BEHAVIOUR

A. The Adaptability of Migratory Behaviour: The Role of Genetic Variances and Covariances

In the last two decades, a number of studies have demonstrated the presence of moderate to high amounts of additive genetic variation in migratory traits in the laboratory and in the wild. Mean heritabilities for migratory traits obtained under laboratory conditions are in accordance with estimates obtained in the wild ($h^2 = 0.40$ and 0.45, respectively), and also there is currently no indication that phenotypic variation is lower under controlled laboratory conditions (see review by Pulido and Berthold, 2003). Therefore, experimental studies may help us to predict evolutionary responses of migratory behaviour to natural selection in the wild.

Most studies on the control and the evolution of migration have been conducted in the blackcap (*Sylvia atricapilla*), the model species for nocturnal passerine migrants. In a series of common-garden, crossbreeding and selection experiments it has been shown that this species has an extraordinarily high adaptability, and a high potential for evolutionary change (Berthold, 1998b; Pulido, 2000). In an artificial selection experiment, for instance, the onset of autumn migratory activity was delayed by almost 2 weeks after two generations of directional artificial selection (see Figure 2; Pulido *et al.*, 2001a). Three to six generations of directional selection on migratoriness transformed a partially migratory blackcap population into a sedentary or completely migratory population (Berthold *et al.*, 1990). These strong selection responses were in accord with responses predicted from the amount of genetic variation found in these populations (Figure 3; Pulido *et al.*, 1996, 2001a). Recent selection experiments have further demonstrated that other adaptive changes that are expected as a response to climatic warming, like the decrease in migration distance and the evolution of sedentariness in migrants can very rapidly be achieved in a few generations (F. Pulido and P. Berthold, unpubl.; see Pulido and Berthold, 2003).

As migratory traits are part of a syndrome, i.e., a suite of co-adapted traits, we expect genetic correlations to be major determinants of evolutionary trajectories and of the rate of adaptive evolution (Dingle, 1996). Presently, genetic correlations have only been estimated for migratory behaviour in the blackcap (Pulido *et al.*, 1996; Pulido and Berthold, 1998; Pulido, 2000), but similar phenotypic correlations among migratory traits in other species suggest that these results may be of general validity (Pulido and Berthold, 2003). In the blackcap, moderate to high genetic correlations among migratory traits, i.e., incidence, intensity and timing of migratory activity, suggest that selection on

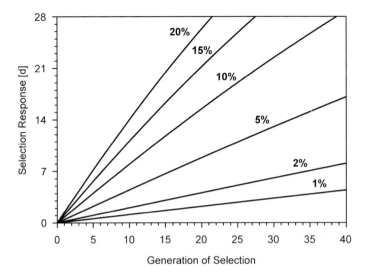

Figure 2   Expected response to directional selection for later onset of migratory activity in southern German blackcaps. Selection responses were estimated from mean phenotypic and genetic variation of this trait in this population (see Pulido *et al.*, 2001a,b). Lines give selection responses under different selection intensities. Percentages above lines give selection intensity as the proportion of individuals that were not allowed to reproduce (from Pulido and Berthold, 2003).

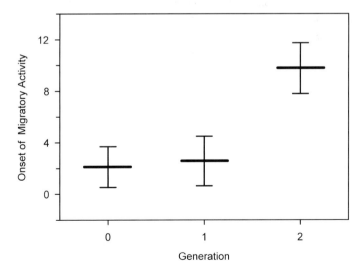

Figure 3   Observed response of the residual onset of migratory activity to two generations of artificial selection for later onset of migratory activity in southern German blackcaps (from Pulido *et al.*, 2001a). Bars give standard errors.

one migratory trait will strongly influence many other traits. Extant genetic correlations among migratory traits may facilitate rapid adaptation to changes in breeding and wintering latitude or altitude, and to phenological changes. The correspondence between the direction of selection vectors and genetic correlations is most probably the result of recurrent and persistent environmental changes in the evolutionary history of migratory bird populations (Pulido and Berthold, 1998). But not all correlations are in the expected direction of selection vectors. Recent studies on the relationship between hatching date, timing of migration and timing of moult suggest that genetic correlations may prevent or delay independent adaptive evolution of timing of breeding, juvenile moult and migration (Coppack et al., 2001; Pulido and Coppack, 2004; Coppack and Pulido, 2004, this volume). We currently do not know whether this will impose serious constraints on the adaptability of birds to global warming. Although genetic correlations have been shown to strongly determine evolutionary trajectories during a relatively long period of time (see Björklund, 1996; Schluter, 1996, 2000), recent experimental studies suggest that rapid response to multivariate selection may be possible despite antagonistic genetic correlations (Beldade et al., 2002). Intraspecific variation in variance–covariance matrices (i.e., G-matrices) in birds (e.g., Badyaev and Hill, 2000; Roff et al., 2004) may be an indication that genetic correlations may be adaptively modified in a short time (see discussion on the evolution of reaction norms in Coppack and Pulido, 2004, this volume).

B. Constraints on Adaptive Evolution of Migration

At present, there are only two studies that identified potential constraints on adaptive evolution of migratory behaviour due to the lack of additive genetic variation on migratory traits. In a field study on Spanish pied flycatchers *F. hypoleuca*, no evidence for genetic variation in the timing of spring arrival was found (Potti, 1998). In a common-garden experiment on European blackbirds *Turdus merula*, a corresponding result was obtained for the amount of migratory activity, which is equivalent to migration distance in the wild (Partecke, 2002). We do not know why in these particular studies no significant heritabilities of migratory traits were found. One possible explanation is that the traits under scrutiny were difficult to measure and that measurement error may have inflated phenotypic variation (Pulido and Berthold, 2003). Alternatively, low heritabilities could result from increased sensitivity to environmental variation or strong genotype–environment interaction (e.g., Stirling et al., 2002), and this may be adaptive. This has been discussed for the termination of migration in the blackcap and other species (Pulido, 2000). Other explanations, like the erosion of genetic variation by natural selection or bottlenecks, need to be further explored.

Migratory birds, and long-distance migrants in particular, may have problems to adapt because of their complex life cycles and their vulnerability to changes in different areas (Coppack and Both, 2002). In fact, there are a number of examples of migratory species that have undergone population crashes as a consequence of rare climatic events on the wintering area, e.g., droughts (e.g., Cavé, 1983; Jones, 1987; Peach et al., 1991; Szép, 1995), and in the breeding area, e.g., due to cold spells in spring or autumn (e.g., Brown and Brown, 1998, 1999, 2000). Milder conditions on the breeding grounds in winter might aggravate the situation for migratory birds, as survival of resident bird species and short distance migrants is likely to increase under these conditions (Forsman and Mönkkönen, 2003; Lemoine and Böhning-Gaese, 2003). This will lead to increased competition within and among species. As a consequence of the competition of long-distance migrants with an increasing number of birds that do not migrate, or that winter close to the breeding grounds, long-distance migrants are expected and have been shown to decrease in numbers (Berthold, 1991; Berthold et al., 1998; Lemoine and Böhning-Gaese, 2003). Moreover, long-distance migrants may take longer to adapt to changing conditions on the breeding grounds than short-distance migrants because they are less pheno-typically flexible and/or may not be exposed to the "climatic information", like the NAO, that is correlated with conditions on the breeding grounds (Butler, 2003; Hüppop and Hüppop, 2003). Moreover, differential changes in phenology *en route* may constrain adaptation to the advancement of spring on the breeding grounds, i.e., birds may be unable to arrive earlier on the breeding grounds because conditions at stopover sites do not allow earlier migration (Strodde, 2003).

C. The Study of Adaptive Change in Migratory Birds

A number of recent changes in migratory behaviour have been attributed to climate change. Increasing number of residents, decreasing migration distances, later departure from and earlier return to the breeding grounds, and the change of migratory direction as a consequence of the establishment of new wintering areas have been the trends reported for numerous bird species (reviewed by Berthold, 1998a, 2001; Fiedler, 2003). These changes in migratory behaviour in response to global warming are characterised by a general trend for reduced migratoriness which may be a consequence of increasing survival probabilities at or near the breeding grounds outside the reproductive seasons (Berthold,1998a; Pulido and Berthold, 1998; Coppack and Both, 2002).

There have basically been three approaches used to detect and study these trends: population monitoring (trapping, banding, song recording), population studies with individually marked birds and experimental studies including common-garden and selective breeding experiments (see Fiedler and Pulido, 2004). Monitoring studies have provided the largest body of evidence for phenotypic changes in migratory behaviour, but only few have been able to investigate the causes of these changes (Fiedler, 2003; Møller and Merilä, 2004, this volume). Long-term population studies are the only means by which fitness differences among different migration strategies can be assessed. Moreover, population studies have helped to assess the amount of additive genetic variation and covariation present in the wild (e.g., Potti, 1998; Møller, 2001; reviewed by Pulido and Berthold, 2003). However, in contrast to avian morphology and breeding biology (see above), migratory behaviour has rarely been the focus of investigation in long-term population studies. The main reasons are that most model species in long-term population studies are non-migratory (e.g., tits), and that reliable measurements of migratory traits are difficult to obtain. Moreover, such studies require high recovery probabilities both on the breeding, and—for studies on migration distance and direction—on the wintering grounds, which are not given in small bird species (Fiedler and Pulido, 2004).

D. Phenotypic Changes in Migratory Behaviour and Their Potential Causes

Apart from changes in laying date, shifts in the timing of migration, spring arrival in particular, are probably the most frequently reported and best investigated responses of avian migratory behaviour to recent climatic change (e.g., Gatter, 1992; Mason, 1995; Sparks, 1999; Forchhammer et al., 2002; Tryjanowski et al., 2002; Butler, 2003; Cotton, 2003; Hüppop and Hüppop, 2003; Jenni and Kéry, 2003; Lehikoinen et al., 2004, this volume). Although these monitoring studies have revealed patterns of change, hitherto, only one study has discussed the potential mechanism underlying these changes. In their analyses of long-term trapping data from the island of Heligoland, Hüppop and Hüppop (2003) found a trend for earlier spring migration in 23 of 24 species. Among-year-variation in spring passage was best explained by the NAO in the preceding winter in long-distance migrants, and local temperatures during migration in short-distance migrants. This finding prompted them to conclude that variation in the timing of spring arrival is a consequence of phenotypic flexibility rather than microevolutionary change. Moreover, they hypothesised that this flexible response is an adaptation to large among-year variation in winter and spring temperatures. Correlations of the timing of spring arrival with the NAO (Forchhammer et al., 2002; Sokolov and Kosarev, 2003), with temperatures on migration (Huin and Sparks, 1998, 2000; Sokolov et al., 1998), on the breeding sites (Sparks and Braslavská, 2002; Tryjanowski et al., 2002), and with temperatures and precipitation in Africa (Cotton, 2003; Sokolov and Kosarev, 2003) have been reported for a number of species, including long-distance migrants. Butler (2003), Forchhammer et al. (2002) and

Tryjanowski *et al.* (2002) found that the immediate influence of climate on the timing of spring migration in long-distance migrants is smaller than in short-distance migrants. Migration in this group of birds has been previously considered to be predominately determined by endogenous spatio-temporal programmes that are insensitive to environmental perturbation (Berthold, 1996).

But to what extent does the correlation between migration phenology and temperatures (or climatic indices) provide evidence for assuming that changes have solely occurred as a consequence of adaptive phenotypic plasticity? We think that such a conclusion is only justified if alternative models that include climate variables at the presumed time of selection, e.g., spring temperatures in the preceding year or temperatures on the wintering site, are tested for best fit and can be excluded (e.g., Forchhammer *et al.*, 2002). It is, for instance, conceivable that due to the correlation between climatic conditions on the breeding grounds in spring and in the non-breeding area in winter, climate-induced selection in winter causes an apparent correlation between spring temperatures in the breeding areas (and on migration) and spring arrival (Møller and Merilä, 2004, this volume). For conclusively demonstrating that populational variation in migration phenology, or in any other trait, is a consequence of phenotypic plasticity alone, longitudinal data, i.e., data from individually marked birds measured in more than one year, are required (see Przybylo *et al.*, 2000; Møller, 2002; Sheldon *et al.*, 2003; Saino *et al.*, 2004). By comparing individual response with the population response to changes in temperature it can be assessed whether phenotypic change in the population can be explained by the mean reaction norm in the population. However, it is likely that phenotypic adaptation to climatic change often involves both phenotypic response at the individual level and genetic changes at the population level (Pulido *et al.*, 2001b; Pulido, 2004; see Réale *et al.*, 2003 for an example in a mammal).

A different approach for studying the genetics of migration is monitoring the number of overwintering birds. Fluctuations in the numbers of birds observed in winter in different areas can provide circumstantial evidence for the genetic basis of migratory behaviour and residency. In an analysis of ringing recoveries of stonechats *Saxicola torquata* wintering in Belgium, Dhondt (1983) showed that fluctuations in the number of wintering birds was correlated with winter temperatures in the previous year indicating that "differential survival of overwintering and migratory individuals takes place and that individuals differ in their tendency to migrate". Although Dhondt could not test whether the consistency of individual behaviour was due to inheritance, cultural transmission, imprinting, or to other mechanisms, he concluded that in view of the rapidity of changes in the proportion of migrants a genetic basis of migratory behaviour was "plausible". Assortative mating according to migratory habits—which was found to be strong in this population—could explain the strong response to selection caused by severe winters. Further evidence for assortative mating according to migratory status has been recently provided in the blackcap using stable isotope signatures (S. Bearhop *et al.*, unpubl.). Strong selection in combination with assortative mating may indeed lead to fast rapid evolutionary change, although evidence from ringing studies can at best be circumstantial. However, this potential for fast evolutionary change should be considered when drawing inferences on the causes of phenotypic change only from the rate of phenotypic change. Without any other evidence, long-term phenotypic change cannot be attributed to changes through phenotypic plasticity, as has been done in some recent studies.

Another population study providing evidence for evolutionary changes in migratory behaviour was conducted by Adriaensen *et al.* (1993) on the great crested grebe *Podiceps cristatus*. Their analysis of ring recoveries revealed that the proportion of Dutch grebes wintering in the breeding area had increased from 22% for the period before 1970 to over 80% in the years after 1980. This increase of residents in the Netherlands was paralleled by an overall increase of the number of great crested grebes breeding in northwestern Europe, and a significant decrease in the number of birds wintering in Switzerland. Circumstantial evidence based on the time lag between numbers of overwintering birds in different areas and on migration distances of Dutch grebes hatched at different times of the year supported the hypothesis that this change in the proportion of resident bird was a consequence of evolutionary change. Other studies have tried to explain changes in migratory behaviour in terms of adaptive evolution of inherited traits (e.g., Able and Belthoff, 1998; Hill *et al.*, 1998). These studies, however, could not rigorously test the hypothesis of genetic change.

E. Microevolutionary Change in Migratory Behaviour

Currently, there are three studies that have investigated evolutionary change of migratory behaviour in response to climate change (see also Berthold *et al.*, 1992 for microevolutionary change caused by other selective agents).

Brown and Brown (2000) demonstrated that in a cliff swallow (*Petrochelidon pyrrhonota*) population in central North America a long period of cold weather in spring had caused strong viability selection on the timing of spring arrival. They found that first-capture dates (an indicator of spring arrival) in the years before selection had been significantly later in survivors than in non-survivors, and that in the generation hatched after selection the frequency of birds arriving early in the season had significantly decreased. Brown and Brown (2000) hypothesised that this mortality in years with longer periods of cold weather in spring could mitigate the effects of selection in "normal" years in which birds arriving early in spring have the highest breeding success and produce the offspring with the highest survival. Because of these rare but regularly recurring years with high spring mortality, however, arrival date in this cliff swallow population does not change in the long run. Although, it could not be tested whether this population actually possesses significant amounts of additive genetic variation for this trait, all evidence provided (e.g., significant repeatability) favours this assertion.

This study currently provides the strongest case for microevolutionary change of migratory behaviour in response to an extreme climatic event (Pulido, 2004). Furthermore, this study shows how oscillating selection may slow-down or prevent long-term evolutionary changes (see also Grant and Grant, 2002). In view of the expected increase of climate extremes like droughts and floods (e.g., Easterling *et al.*, 2000a), and a possible reduction of frosts in spring (Scheifinger *et al.*, 2003), this finding may be particularly important for interpreting and predicting long-term microevolutionary changes in phenology.

In another long-term study on a Danish barn swallow population, Møller (2004; Møller and Merilä, 2004, this volume) found a significant trend for earlier spring arrival in the period from 1971 to 2002. This trend was predominantly due to a marked advancement of spring arrival since 1985, and was paralleled by a decrease of estimated breeding values for arrival dates over time, suggesting that phenotypic change was, at least, partly caused by genetic changes in the population. Møller hypothesised that mean arrival data is early when environmental conditions in Northern Africa during migration are unfavourable, because birds migrating late, which presumably are in bad condition, do not arrive on the breeding grounds. In years when conditions at northern African stopover sites are favourable, more birds may survive migration (Møller and Szép, 2002). Thus, a larger number of birds with late arrival should be found in the population.

Although this study could potentially be the first in demonstrating evolutionary change in migratory behaviour in response to climatic change, it is not without problems. The reliability of estimated breeding values depends very much on the influential environmental variables being considered, and the depth and the complexity of the pedigree used. If only little pedigree information is available, estimated breeding values will not differ from phenotypic values (E. Postma, pers. comm.). Because in Møller's analysis only arrival dates of fathers and sons were used, the changes in estimated breeding values may not reflect genetic but rather phenotypic change. Moreover, in northern Italian barn swallows, among-year variation in arrival date was caused by individual phenotypic adjustment of adult birds to environmental conditions in the wintering area (Saino *et al.*, 2004), as is the case for some of the variation in arrival date in Denmark (A.P. Møller, unpubl.). If different populations of the same species have similar mechanisms of adaptation, it is likely that the trend for earlier spring arrival in Danish barn swallows may partly represent adaptation by phenotypic plasticity rather than microevolutionary change.

In contrast to these field studies, we used a common garden experiment to study changes in autumn migration in the blackcap (F. Pulido and P. Berthold, in prep.). From 1988 through 2001, each year (but one) we collected nestlings randomly from a southern German population and transferred them to the laboratory. There, we hand-raised them and kept them under standardised conditions. In autumn, we registered migratory activity of each individual using an identical protocol throughout the study period (see Pulido *et al.*, 2001a; Pulido and Coppack, 2004). Some of these birds were used for breeding experiments, which made it possible to exclude maternal effects and served as a control. By using this experimental protocol, we could show that in the course of 13 years, the amount of migratory activity has dramatically decreased by about one standard deviation. This change in the amount of migratory activity has been primarily caused by a strong reduction of the intensity of migration, i.e., the activity per night, and to a lesser extent as a consequence of a delay of the onset of autumn migration. Changes in the termination of migration were not significant, as was expected from the low heritability of this trait. Variation in migration distance assessed by ringing data is very large and the number of recoveries too low to be able to show this trend in the reduction of migration distance. However, in the last decade, an increasing number of anecdotal observations of blackcaps wintering in central and northern Europe has been reported (e.g., Andres and Bersuder, 1992; Fransson and Stolt, 1994), which may reflect a reduction of migratoriness in these areas. We think that the genetic changes in migratory behaviour found in southern German blackcaps are very likely to result from adaptive response to selection imposed by recent climate change and not from massive gene flow, as neither ringing recoveries nor biometrical data obtained from trapping programmes show a tendency for an influx of individuals from other populations. Moreover, the changes observed are in accord with the changes predicted under a global-warming scenario (see above, and Pulido and Berthold, 1998). Unfortunately, we currently have no field data on survival and reproduction that would allow us to test whether the observed changes are actually in accord with predictions from selection intensities and heritabilities.

Using different approaches, the three studies described suggest that migratory behaviour may readily respond to selection imposed by climatic change. They support previous findings reporting high potential for adaptive change of migratory behaviour in birds (reviewed by Berthold, 1998b; Pulido and Berthold, 2003; Pulido, 2004). The study by Møller (2004, unpubl.) is an example of how ecological, demographic and pedigree data could be used to study the causes of phenotypic change in the timing of spring arrival. The experimental studies in the blackcap highlight the role of genetic variation and covariation for reconstructing selection events. Studies combining both approaches, using long and complex pedigrees (as available for many hole-nesting birds), and extending the set of characters under scrutiny to other life-cycle stages (see Coppack and Pulido, 2004, this volume) are urgently needed to understand the adaptation and evolution of life cycles. This will probably only be possible by incorporating new methods (e.g., stable isotopes, genetic markers, satellite tracking; cf. Fiedler and Pulido, 2004) into the research agenda, and by combining the efforts of several research groups to study one or a few model system(s).

VIII. CONCLUSIONS AND OUTLOOK

Birds may very rapidly respond to changing environmental conditions imposed by global climatic change. Although fossil records in other taxa suggest that the most likely response to climatic change is range shift rather than adaptive evolution (e.g., Cronin and Schneider, 1990; Parmesan *et al.*, 2000), the validity of this inference for the current evolutionary response is questionable (Travis and Futuyma, 1993). Adaptability to temperature changes is likely to have evolved by correlational selection, as a result of environmental fluctuation during the evolutionary history of many bird populations (Burton, 1995). High levels of genetic variation, favourable genetic correlations, cogradient variation and adaptive plasticity may allow rapid responses to current environmental changes (Pulido, 2000, 2004; Pulido and Berthold, 2003; Coppack *et al.*, 2003). Moreover, unidirectional gene flow from populations or species better adapted to warmer environmental conditions may further accelerate adaptive evolution. If climate change persists and is accompanied by an increase of climatic variability—as currently predicted (Houghton *et al.*, 2001)—the rate of evolutionary change may be much lower than found in studies on single selection events, because of oscillating selection and the erosion of genetic variation. In addition, environmental variation may not be buffered by phenotypic plasticity, or may become maladaptive (Schlaepfer *et al.*, 2002; Coppack and Pulido, 2004, this volume), which may further accelerate the erosion of genetic variation. Unfavourable genetic correlations are likely to constrain the rate of adaptive evolution (Etterson and Shaw, 2001), but adaptive changes in the variance–covariance matrix could evolve if the direction of selection remains constant over a longer period of time (Endler, 1995). Alternatively, genetic covariances may not change, probably because they are caused by pleiotropic gene effects. Then, unfavourable genetic covariation could determine evolutionary trajectories over a long period of time (Schluter, 1996, 2000).

A complete understanding of adaptive evolution will only be achieved if ecological and genetic studies are integrated, and the influences of the environment, different fitness components, phenotypic plasticity, genetic covariation and gene flow can be evaluated at the same time. Recent studies have followed such an integrative approach, and have revealed that adaptive evolution is determined by a variety of factors and their complex interactions (see, for instance, Sheldon *et al.*, 2003). Data from other long-term population studies should be used to reconstruct adaptive changes in the same way, and to unravel among-population differences in evolutionary response (e.g., using the data on blue and great tits analysed by Visser *et al.*, 2003). Such an approach will help us to reveal general patterns of adaptation, and to study changes in the adaptability of particular traits and populations in response to changes in the environment.

ACKNOWLEDGEMENTS

We thank Timothy Coppack, Anders Møller, and two anonymous reviewers for critically reading the manuscript.

REFERENCES

Able, K. and Belthoff, J.R. (1998) *Proc. R. Soc. Lond. B* **265**, 2063–2071.
Adriaensen, F., Ulenaers, P. and Dhondt, A.A. (1993) *Ardea* **81**, 59–70.
Andres, C. and Bersuder, D. (1992) *Ciconia* **16**, 131–135.
Arnold, S.J. (1988) In: *Proceedings of the Second International Conference on Quantitative Genetics* (Ed. by B.S. Weir, E.J. Eisen, M.M. Goodman and G. Namkoong), pp. 619–636. Sinauer, Sunderland, MA.
Arnold, S.J. (1994) In: *Quantitative Genetic Studies of Behavioral Evolution* (Ed. by C.R.B. Boake), pp. 17–48. University of Chicago Press, Chicago.
Badyaev, A.V. and Hill, G.E. (2000) *Evolution* **54**, 1784–1794.
Barnosky, A.D., Hadly, E.A. and Bell, C.J. (2003) *J. Mammal.* **84**, 354–368.
Barrett, S.C.H. and Charlesworth, D. (1991) *Nature* **352**, 522–524.
Bauer, H.-G. and Berthold, P. (1997) *Die Brutvögel Mitteleuropas. Bestand und Gefährdung*, 2nd Ed., Aula, Wiesbaden.
Beldade, P., Koops, K. and Brakefield, P.M. (2002) *Nature* **416**, 844–847.
Belliure, J., Sorci, G., Møller, A.P. and Clobert, J. (2000) *J. Evol. Biol.* **13**, 480–487.
Benning, T.L., LaPointe, D., Atkinson, C.T. and Vitousek, P.M. (2002) *Proc. Natl. Acad. Sci. USA* **99**, 14246–14249.
Berthold, P. (1990) *Verh. Dtsch. Zool. Ges.* **83**, 227–244.
Berthold, P. (1991) *Acta XX Congr. Internat. Ornithol.*, 780–786.
Berthold, P. (1996) *Control of Bird Migration*. Chapman and Hall, London.
Berthold, P. (1998a) *Naturw. Rundsch.* **51**, 337–346.
Berthold, P. (1998b) *Zoology* **101**, 235–245.
Berthold, P. (2001) *Bird Migration*, 2nd Ed., Oxford University Press, Oxford.
Berthold, P., Mohr, G. and Querner, U. (1990) *J. Ornithol.* **131**, 33–45.
Berthold, P., Helbig, A.J., Mohr, G. and Querner, U. (1992) *Nature* **360**, 668–670.
Berthold, P., Helbig, A.J., Mohr, G., Pulido, F. and Querner, U. (1996) *Max-Planck-Gesellschaft Jahrbuch* **1996**, 346–354.

Berthold, P., Fiedler, W., Schlenker, R. and Querner, U. (1998) *Naturwiss* **85**, 350–353.

Bijlsma, R., Bundgaard, J. and Boerema, A.C. (2000) *J. Evol. Biol.* **13**, 502–514.

Björklund, M. (1996) *Evol. Ecol.* **10**, 423–431.

Boag, P.T. and van Noordwijk, A.J. (1987) In: *Avian Genetics* (Ed. by P.A. Buckley and F. Cooke), pp. 45–78. Academic Press, London.

Boake, C.R.B., Arnold, S.J., Breden, F., Meffert, L.M., Ritchie, M.G., Taylor, B.J., Wolf, J.B. and Moore, A.J. (2002) *Am. Nat.* **160**, S143–S159.

Booy, G., Hendriks, R.J.J., Smulders, M.J.M., Van Groenendael, J.M. and Vosman, B. (2000) *Plant Biol.* **2**, 379–395.

Both, C. and Visser, M.E. (2001) *Nature* **411**, 296–298.

Bradshaw, A.D. (1991) *Phil. Trans. R. Soc. Lond. B* **333**, 289–305.

Bradshaw, W.E. and Holzapfel, C.M. (2001) *Proc. Natl. Acad. Sci. USA* **98**, 14509–14511.

Bradshaw, A.D. and McNeilly, T. (1991) *Ann. Bot.* **67(suppl. 1)**, 5–14.

Briskie, J.V. and Mackintosh, M. (2004) *Proc. Natl. Acad. Sci. USA* **101**, 558–561.

Brown, C.R. and Brown, M.B. (1998) *Evolution* **52**, 1461–1475.

Brown, C.R. and Brown, M.B. (1999) *Ibis* **141**, 652–659.

Brown, C.R. and Brown, M.B. (2000) *Behav. Ecol Sociobiol.* **47**, 339–345.

Brown, J.L., Li, S.-H. and Bhagabati, N. (1999) *Proc. Natl. Acad. Sci. USA* **96**, 5565–5569.

Bumpus, H.C. (1899) *Biological Lectures of the Woods Hole Marine Biological Station* **6**, 209–226.

Burton, J.F. (1995) *Birds and Climate Change.* Christopher Helm, London.

Butler, C.J. (2003) *Ibis* **145**, 484–495.

Cavé, A.J. (1983) *Ardea* **71**, 217–224.

Conner, J.K. (2003) *Ecology* **84**, 1650–1660.

Conover, D.O. and Schultz, E.T. (1995) *Trends Ecol. Evol.* **10**, 248–252.

Cotton, P.A. (2003) *Proc. Natl. Acad. Sci. USA* **100**, 12219–12222.

Coppack, T. and Both, C. (2002) *Ardea* **90**, 369–378.

Coppack, T., Pulido, F. and Berthold, P. (2001) *Oecologia* **128**, 181–186.

Coppack, T., Pulido, F., Czisch, M., Auer, D.P. and Berthold, P. (2003) *Proc. R. Soc. Lond. B* **270(suppl.)**, S43–S46.

Crick, H.Q.P. and Sparks, T.H. (1999) *Nature* **399**, 423–424.

Crick, H.Q.P., Dudley, C., Glue, D.E. and Thompson, D.L. (1997) *Nature* **388**, 526.

Cronin, T.M. and Schneider, C.E. (1990) *Trends Ecol. Evol.* **5**, 275–279.

Daniels, S.J. and Walters, J.R. (2000) *Condor* **102**, 482–491.

DeWitt, T.J., Sih, A. and Wilson, D.S. (1998) *Trends Ecol. Evol.* **13**, 77–81.

Dhondt, A.A. (1983) *Ring. Migr.* **4**, 155–158.

Dingle, H. (1996) *Migration. The Biology of Life on the Move.* Oxford University Press, Oxford.

Dunham, J., Peacock, M., Tracy, C.R., Nielsen, J. and Vinyard, G. (1999) *Cons. Ecol.* **3**, 2 [online]. Available from the Internet. URL: http://www.consecol.org/vol3/iss1/art2

Dunn, P.O. and Winkler, D.W. (1999) *Proc. R. Soc. Lond. B* **266**, 2487–2490.

Easterling, D.R., Evans, J.L., Groisman, P.Ya., Karl, T.R., Kunkel, K.E. and Ambenje, P. (2000a) *Bull. Am. Meteorol. Soc.* **81**, 417–425.

Easterling, D.R., Meehl, G.A., Parmesan, C., Changon, S.A., Karl, T.R. and Mearns, L.O. (2000b) *Science* **289**, 2068–2074.

Edmands, S. and Timmerman, C.C. (2003) *Conserv. Biol.* **17**, 883–892.

Endler, J.A. (1986) *Natural Selection in the Wild*. Princeton University Press, Princeton, NJ.

Endler, J.A. (1995) *Trends Ecol. Evol.* **10**, 22–29.

Etterson, J.R. and Shaw, R.G. (2001) *Science* **294**, 151–154.

Ewen, F., Cassey, P. and Møller, A.P. (2004) *Proc. R. Soc. Lond. B* **271**, 7277–7282.

Faivre, B., Secondi, J., Ferry, C., Chastragnant, L. and Cézilly, F. (1999) *J. Avian Biol.* **30**, 152–158.

Falconer, D.S. and Mackay, T.F.C. (1996) *Introduction to Quantitative Genetics*, 4th Ed., Longman, Harlow.

Fiedler, W. (2003) In: *Avian Migration* (Ed. by P. Berthold, E. Gwinner and E. Sonnenschein), pp. 21–38. Springer, Berlin.

Fiedler, W. and Pulido, F. (2004) In: *Proceedings of the 23rd International Ornithological Congress* (Ed. by R. Schodde), Acta Zoologica Sinica.

Flux, J.E.C. and Flux, M.M. (1982) *Naturwiss* **69**, 96–97.

Forchhammer, M.C., Post, E. and Stenseth, N.C. (2002) *J. Anim. Ecol.* **71**, 1002–1014.

Forsman, J.T. and Mönkkönen, M. (2003) *J. Biogeogr.* **30**, 55–70.

Forstmeier, W., Leisler, B. and Kempenaers, B. (2001) *Proc. R. Soc. Lond. B* **268**, 1583–1588.

Fraedrich, K., Gerstengarbe, F.-W. and Werner, P.C. (2001) *Climate Change* **50**, 405–417.

Frankham, R. (1995a) *Conserv. Biol.* **9**, 792–799.

Frankham, R. (1995b) *Genet. Res. Camb.* **66**, 95–107.

Frankham, R., Ballou, J.D. and Briscoe, D.A. (2002) *Introduction to Conservation Genetics*. Cambridge University Press, Cambridge.

Fransson, T. and Stolt, B.-O. (1994) *Ornis Svecica* **4**, 105–112.

Futuyma, D.J. (1998) *Evolutionary Biology*, 3rd Ed., Sinauer, Sunderland, MA.

Gaggiotti, O.E. (2003) *Ann. Zool. Fennici* **40**, 155–168.

Gatter, W. (1992) *J. Ornithol.* **133**, 427–436.

Geber, M.A. and Dawson, T.E. (1993) In: *Biotic Interactions and Global Change* (Ed. by P.M. Kareiva, J.G. Kingsolver and R.B. Huey), pp. 179–197. Sinauer, Sunderland, MA.

Gibbs, J.P. (2001) *Biol. Conserv.* **100**, 15–20.

Gomulkiewicz, R. and Holt, R.D. (1995) *Evolution* **49**, 201–207.

Grant, B.R. and Grant, P.R. (1989) *Evolutionary Dynamics of a Natural Population. The Large Cactus Finch of the Galápagos*. University of Chicago Press, Chicago.

Grant, P.R. and Grant, B.R. (1992) *Science* **256**, 193–197.

Grant, P.R. and Grant, B.R. (1994) *Evolution* **48**, 297–316.

Grant, P.R. and Grant, B.R. (1995) *Evolution* **49**, 241–251.

Grant, B.R. and Grant, P.R. (1996) *Ecology* **72**, 500–509.

Grant, P.R. and Grant, B.R. (2002) *Science* **296**, 707–711.

Grant, B.R. and Grant, P.R. (1993) *Proc. R. Soc. Lond. B* **251**, 111–117.

Grieco, F., van Noordwijk, A.J. and Visser, M.E. (2002) *Science* **296**, 136–138.

Grosch, K. (2003) *Evol. Ecol.* **17**, 1–17.

Hasselquist, D., Wasson, M.F. and Winkler, D.W. (2001) *Behav. Ecol.* **12**, 93–97.

Henderson, C.R. (1986) *J. Anim. Sci.* **63**, 208–216.

Hendry, A.P. and Kinnison, M.T. (1999) *Evolution* **53**, 1637–1653.

Hill, W.G. and Caballero, A. (1992) *Ann. Rev. Ecol. Syst.* **23**, 287–310.

Hill, G.E., Sargent, R.R. and Sargent, M.B. (1998) *Auk* **115**, 240–245.

Hoffmann, A.A. and Blows, M.W. (1993) In: *Biotic Interactions and Global Change* (Ed. by P.M. Kareiva, J.G. Kingsolver and R.B. Huey), pp. 165–178. Sinauer, Sunderland, MA.

Hoffmann, A.A. and Parsons, P.A. (1997) *Extreme Environmental Change and Evolution.* Cambridge University Press, Cambridge.

Hoffmann, A.A., Hallas, R.J., Dean, J.A. and Schiffer, M. (2003) *Science* **301**, 100–102.

Holt, R.D. (1990) *Trends Ecol. Evol.* **5**, 311–315.

Houghton, J.T., Ding, D.J., Nouger, M., van der Linden, P.J. and Xiaosu, D. (2001) *Contribution of Working Group I to the Third Assessment Report of the Intergovernmental Panel on Climate Change (IPCC).* Cambridge University Press, Cambridge.

Hughes, L. (2000) *Trends Ecol. Evol.* **15**, 56–61.

Huin, N. and Sparks, T.H. (1998) *Bird Study* **45**, 361–370.

Huin, N. and Sparks, T.H. (2000) *Bird Study* **47**, 22–31.

Huntley, B. (1994) *Ibis* **137**, S127–S138.

Hüppop, O. and Hüppop, K. (2003) *Proc. R. Soc. Lond. B* **270**, 233–240.

Jakober, H. and Stauber, W. (2000) *J. Ornithol.* **141**, 408–417.

Jenni, L. and Kéry, M. (2003) *Proc. R. Soc. Lond. B* **270**, 1467–1471.

Johnston, R.F., Niles, D.M. and Rohwer, S.A. (1972) *Evolution* **26**, 20–31.

Jones, G. (1987) *Ibis* **129**, 274–280.

Juillard, R., McCleery, R.H., Clobert, J. and Perrins, C.M. (1997) *Ecology* **78**, 394–404.

Keller, L. and Waller, D.M. (2002) *Trends Ecol. Evol.* **17**, 230–241.

Keller, L.F., Arcese, P., Smith, J.N.M., Hochachka, W.M. and Stearns, S. (1994) *Nature* **372**, 356–357.

Keller, L.F., Grant, P.R., Grant, B.R. and Petren, K. (2002) *Evolution* **56**, 1229–1239.

Kempenaers, B., Adriaensen, F., van Noordwijk, A.J. and Dhondt, A.A. (1996) *Proc. R. Soc. Lond. B* **263**, 179–185.

Kingsolver, J.G., Hoekstra, H.E., Hoekstra, J.M., Berrigan, D., Vignieri, S.N., Hill, C.E., Hoang, A., Gibert, P. and Beerli, P. (2001) *Am. Nat.* **157**, 245–261.

Kinnison, M.T. and Hendry, A.P. (2001) *Genetica* **112–113**, 145–164.

Kirstensen, T.N., Dahlgaard, J. and Loeschcke, V. (2003) *Conserv. Genet.* **4**, 453–465.

Knott, S.A., Sibly, R.M., Smith, R.H. and Møller, H. (1995) *Funct. Ecol.* **9**, 122–126.

Koike, S. and Higuchi, H. (2002) *Ibis* **144**, 150–152.

Kozlov, M.V. and Berlina, N.G. (2002) *Climate Change* **54**, 387–398.

Kruuk, L.E.B. (2004) *Phil. Trans. R. Soc. Lond. B* **359**, 873–890.

Kruuk, L.E.B., Merilä, J. and Sheldon, B.C. (2001) *Am. Nat.* **158**, 557–571.

Kruuk, L.E.B., Slate, J., Pemberton, J.M., Brotherstone, S., Guinness, F. and Clutton-Brock, T. (2002) *Evolution* **56**, 1683–1695.

Kruuk, L.E.B., Merilä, J. and Sheldon, B.C. (2003) *Trends Ecol. Evol.* **18**, 207–209.

Lacy, R.C. (1987) *Conserv. Biol.* **1**, 143–158.

Lande, R. and Arnold, S.J. (1983) *Evolution* **37**, 1210–1226.

Lande, R. and Shannon, S. (1996) *Evolution* **50**, 434–437.

Leisler, B., Winkler, H. and Wink, M. (2002) *Auk* **119**, 379–390.

Lemoine, N. and Böhning-Gaese, K. (2003) *Conserv. Biol.* **17**, 577–586.

Lenormand, T. (2002) *Trends Ecol. Evol.* **17**, 183–189.

Lesica, P. and Allendorf, F.W. (1995) *Conserv. Biol.* **9**, 753–760.

Ludwichowski, I. (1997) *Vogelwarte* **39**, 103–116.

Lynch, M. and Lande, R. (1993) In: *Biotic Interactions and Global Change* (Ed. by P.M. Kareiva, J.G. Kingsolver and R.B. Huey), Sinauer, Sunderland, MA.

Mason, C.F. (1995) *Bird Study* **42**, 182–189.

McCarty, J.P. (2001) *Conserv. Biol.* **15**, 320–331.

McLaughlin, J.F., Hellmann, J.J., Boggs, C.L. and Ehrlich, P.R. (2002) *Proc. Natl. Acad. Sci. USA* **99**, 6070–6074.

Meehl, G.A., Zwiers, F., Evans, J., Knutson, T., Mearns, L. and Whetton, P. (2000) *Bull. Am. Meteorol. Soc.* **81**, 427–436.

Menzel, A. (2000) *Int. J. Biometeorol.* **44**, 76–81.

Menzel, A. and Fabian, P. (1999) *Nature* **397**, 659.

Merilä, J., Kruuk, L.E.B. and Sheldon, B.C. (2001a) *Nature* **412**, 76–79.

Merilä, J., Kruuk, L.E.B. and Sheldon, B.C. (2001b) *J. Evol. Biol.* **14**, 918–929.

Merilä, J., Sheldon, B.C. and Kruuk, L.E.B. (2001c) *Genetica* **112–113**, 199–222.

Meyers, L.A. and Bull, J.J. (2002) *Trends Ecol. Evol.* **17**, 551–557.

Møller, A.P. (2001) *Proc. R. Soc. Lond. B.* **268**, 203–206.

Møller, A.P. (2002) *J. Anim. Ecol.* **71**, 201–210.

Møller, A.P. and Szép, T. (2002) *Ecology* **83**, 2220–2228.

Møller, A.P. (2004) *Global Change Biol.* (in press).

Mousseau, T.A. and Fox, C.W. (Eds.) (1998) *Maternal Effects as Adaptations*. Oxford University Press, New York.

Mousseau, T.A. and Roff, D.A. (1987) *Heredity* **59**, 181–197.

Nunney, L. (1995) *Evolution* **49**, 389–392.

Nunney, L. (2003) *Ann. Zool. Fenn.* **40**, 185–194.

Paradis, E., Baillie, S.R., Sutherland, W.J. and Gregory, R.D. (1998) *J. Anim. Ecol.* **67**, 518–536.

Parmesan, C. (1996) *Nature* **382**, 765–766.

Parmesan, C. and Yohe, G. (2003) *Nature* **421**, 37–42.

Parmesan, C., Ryrholm, N., Stefanescu, C., Hillk, J.K., Thomas, C.D., Descimon, H., Huntley, B., Kaila, L., KullbergI, J., Tammaru, T., Tennent, W.J., Thomas, J.A. and Warren, M. (1999) *Nature* **399**, 579–583.

Parmesan, C., Root, T.L. and Willig, M.R. (2000) *Bull. Am. Meteorol. Soc.* **81**, 443–450.

Parsons, P.A. (1990) *Trends Ecol. Evol.* **5**, 315–317.

Partecke, J. (2002) Annual cycles of urban and forest-living European Blackbirds (*Turdus merula*): genetic differences or phenotypic plasticity. Ph.D. Thesis. University of Munich, Germany.

Peach, W., Baillie, S. and Underhill, L. (1991) *Ibis* **122**, 300–305.

Pearson, R. (1978) *Climate and Evolution*. Academic Press, London.

Pigliucci, M. (2001) *Phenotypic Plasticity. Beyond Nature and Nurture.* Johns Hopkins University Press, Baltimore, MD.

Piersma, T. and Drent, J. (2003) *Trends Ecol. Evol.* **18**, 228–233.

Potti, J. (1998) *Condor* **100**, 702–708.

Price, T.D. and Boag, P.T. (1987) In: *Avian Genetics* (Ed. by P.A. Buckley and F. Cooke), pp. 257–287. Academic Press, London.

Price, J. and Glick, P. (2001) *The Birdwatcher's Guide to Global Warming.* National Wildlife Federation and American Bird Conservancy, Virginia.

Price, T.D. and Yeh, P. (1999) *Endeavour* **23**, 145–147.

Price, T.D., Brown, C.R. and Brown, M.B. (2000) *Evolution* **54**, 1824–1827.

Price, T.D., Qvarnström, A. and Irwin, D.E. (2003) *Proc. R. Soc. Lond. B* **270**, 1433–1440.

Przybylo, R., Sheldon, B.C. and Merilä, J. (2000) *J. Anim. Ecol.* **69**, 395–403.

Pulido, F. (2000) *Evolutionary Quantitative Genetics of Migratory Restlessness in the Blackcap (Sylvia atricapilla).* Edition Wissenschaft, Reihe Biologie, Vol. 224. Tecctum Verlag, Marburg.

Pulido, F. (2004) In: *Proceedings of the 23rd International Ornithological Congress* (Ed. by R. Schodde), Acta Zoologica Sinica.

Pulido, F. and Berthold, P. (1998) *Biol. Conserv. Fauna* **102**, 206–211.

Pulido, F. and Berthold, P. (2003) In: *Avian Migration* (Ed. by P. Berthold, E. Gwinner and E. Sonnenschein), pp. 53–77. Springer, Berlin.

Pulido, F. and Coppack, T. (2004) *Anim. Behav.* **68**, 167–173.

Pulido, F., Berthold, P. and van Noordwijk, A.J. (1996) *Proc. Natl. Acad. Sci. USA* **93**, 14642–14647.

Pulido, F., Berthold, P., Mohr, G. and Querner, U. (2001a) *Proc. R. Soc. Lond. B* **268**, 953–959.

Pulido, F., Coppack, T. and Berthold, P. (2001b) *Ring* **23**, 149–158.

Randler, C. (2002) *Anim. Behav.* **63**, 103–119.

Rausher, M.D. (1992) *Evolution* **46**, 616–626.

Réale, D., McAdam, A.G., Boutin, S. and Berteaux, D. (2003) *Proc. R. Soc. Lond. B* **270**, 591–596.

Reed, D. and Frankham, R. (2003) *Conserv. Biol.* **17**, 230–237.

Reed, D.H., Lowe, E.H., Briscoe, D.A. and Frankham, R. (2003) *Evolution* **57**, 1822–1828.

Reid, J.M., Arcese, P. and Keller, L.F. (2003) *Proc. R. Soc. Lond. B* **270**, 2151–2157.

Reznick, D.N. and Ghalambor, C. (2001) *Genetica* **112–113**, 183–198.

Reznick, D. and Travis, J. (1996) In: *Adaptation* (Ed. by M.R. Rose and G.V. Lauder), pp. 243–289. Academic Press, San Diego, CA.

Rhymer, J.M. and Simberloff, D.S. (1996) *Ann. Rev. Ecol. Syst.* **27**, 83–109.

Rice, K.J. and Emery, N.C. (2003) *Front. Ecol. Environ.* **1**, 469–478.

Rodríguez-Trelles, F. and Rodríguez, M.A. (1998) *Evol. Ecol.* **12**, 829–838.

Rodríguez-Trelles, F., Rodríguez, M.A. and Scheiner, S.M. (1998) *Cons. Ecol.* **2**, 2, [online]. Available from the Internet. URL: http://www.consecol.org/vol2/iss2/art2.

Roff, D.A. (1997) *Evolutionary Quantitative Genetics.* Chapman and Hall, New York.

Roff, D.A. and Fairbairn, D.J. (1999) *Heredity* **83**, 440–450.

Roff, D.A., Mousseau, T.A., Møller, A.P., de Lope, F. and Saino, N. (2004) *Heredity* **93**, 8–14.

Root, T.L. (1993) In: *Biotic Interactions and Global Change* (Ed. by P.M. Kareiva, J.G. Kingsolver and R.B. Huey), pp. 280–292. Sinauer, Sunderland, MA.

Root, T.L., Price, J.T., Hall, K.R., Schneider, S.H., Rosenzweig, C. and Pounds, J.A. (2003) *Nature* **421**, 57–60.

Safriel, U.N., Volis, S. and Kark, S. (1994) *Isr. J. Plant Sci.* **42**, 331–345.

Saino, N., Szép, T., Romano, M., Rubolini, D., Spina, F. and Møller, A.P. (2004) *Ecol. Lett.* **7**, 21–25.

Scheifinger, H., Menzel, A., Koch, E. and Peter, C. (2003) *Theor. Appl. Climatol.* **74**, 41–51.

Scheiner, S.M. (1993) *Ann. Rev. Ecol. Syst.* **24**, 35–68.

Scheiner, S.M. (2002) *J. Evol. Biol.* **15**, 889–898.

Scheiner, S.M. and Lyman, R.F. (1991) *J. Evol. Biol.* **4**, 23–50.

Schiegg, S., Pasinelli, G., Walters, J.R. and Daniels, S.J. (2002) *Proc. R. Soc. Lond. B* **269**, 1153–1159.

Schlaepfer, M.A., Runge, M.C. and Sherman, P.W. (2002) *Trends Ecol. Evol.* **17**, 474–480.

Schluter, D. (1996) *Evolution* **50**, 1766–1774.

Schluter, D. (2000) *The Ecology of Adaptive Radiation.* Oxford University Press, Oxford.

Sergio, F. (2003) *J. Avian Biol.* **34**, 144–149.

Sheldon, B.C. (1998) *Heredity* **80**, 397–402.

Sheldon, B.C., Kruuk, L.E.B. and Merilä, J. (2003) *Evolution* **57**, 406–420.

Simberloff, D. (1995) *Ibis* **137**, S105–S111.

Slatkin, M. (1987) *Science* **236**, 787–792.

Sokolov, L.V. and Kosarev, V.V. (2003) *Proc. Zool. Inst. Russ. Acad. Sci.* **299**, 141–154.

Sokolov, L.V., Markovets, M.Y., Shapoval, A.P. and Morozov, Y.G. (1998) *Avian Ecol. Behav.* **1**, 1–21.

Sparks, T.H. (1999) *Int. J. Biometeorol.* **42**, 134–138.

Sparks, T.H. and Menzel, A. (2002) *Int. J. Climatol.* **22**, 1715–1725.

Sparks, T.H. and Braslavská, O. (2001) *Int. J. Biometeorol.* **45**, 212–216.

Spottiswoode, C. and Møller, A.P. (2004) *Proc. R. Soc. Lond. B* **271**, 267–272.

Stirling, D.G., Réale, D. and Roff, D.A. (2002) *J. Evol. Biol.* **15**, 277–289.

Storfer, A. (1999) *Biol. Conserv.* **87**, 173–180.

Strodde, P.K. (2003) *Global Change Biol.* **9**, 1137–1144.

Szép, T. (1995) *Ibis* **137**, 162–168.

Templeton, A.R. and Read, B. (1984) *Zoo Biol.* **3**, 177–199.

Thomas, C.D. and Lennon, J.J. (1999) *Nature* **399**, 213.

Travis, J.M.J. (2003) *Proc. R. Soc. Lond. B* **270**, 467–473.

Travis, J. and Futuyma, D.J. (1993) In: *Biotic Interactions and Global Change* (Ed. by P.M. Kareiva, J.G. Kingsolver and R.B. Huey), pp. 251–263. Sinauer, Sunderland, MA.

Trussel, G.C. and Etter, R.J. (2001) *Genetica* **112–113**, 321–337.

Tryjanowski, P., Kuzniak, S. and Sparks, T. (2002) *Ibis* **144**, 62–68.

Valiela, I. and Bowen, J.L. (2003) *Ambio* **32**, 476–480.

van Noordwijk, A.J. and Müller, C.B. (1994) *Physiol. Ecol. (Japan), Spec. Ed.* **29**, 180–194.

Visser, M.E. and Holleman, L.J.M. (2001) *Proc. R. Soc. Lond. B* **268**, 289–294.

Visser, M.E., Adriaensen, F., van Balen, J.H., Blondel, J., Dhondt, A.A., van Dongen, S., du Feu, C., Ivankina, E.V., Kerimov, A.B., de Laet, J., Matthysen, E., McCleery, R., Orell, M. and Thomson, D.L. (2003) *Proc. R. Soc. Lond. B* **270**, 367–372.

Walther, G.-R., Post, E., Convey, P., Menzel, A., Parmesan, C., Beebee, T.J.C., Fromentin, J.-M., Høgh-Guldberg, O. and Bairlein, F. (2002) *Nature* **416**, 389–395.

West-Eberhard, M.J. (2003) *Developmental Plasticity and Evolution.* Oxford University Press, Oxford.

Winkel, W. and Hudde, H. (1996) *J. Orn.* **137**, 193–202.

Winkel, W. and Hudde, H. (1997) *J. Avian Biol.* **28**, 187–190.

Wirtz, P. (1999) *Anim. Behav.* **58**, 1–12.

Yom-Tov, Y. (2001) *Proc. R. Soc. Lond. B* **268**, 947–952.

Yoo, B.H. (1980) *Genet. Res.* **35**, 1–17.

Climate Influences on Avian Population Dynamics

BERNT-ERIK SÆTHER,* WILLIAM J. SUTHERLAND
AND STEINAR ENGEN

I. SUMMARY

Predictions of the consequences of the expected changes in climate on bird population dynamics require a detailed mechanistic understanding. Basically, two different hypotheses have been proposed to explain under which period of the year a change in climate will have the strongest effect on fluctuations in population size. The *tub-hypothesis* proposes fluctuations in population size to be closely related to climate variation during the non-breeding season because in combination with density dependence, the weather conditions determine the number of birds surviving during this critical period of the year. The *tap-hypothesis* predicts annual variation in population size to be related to the weather during the breeding season because this will influence the inflow of new recruits into the population the following year. We examine the validity of these hypotheses by reviewing studies that have related fluctuations in population sizes to local weather variables or to large-scale climate patterns such as the North Atlantic Oscillation (NAO). We find the tub-hypothesis to be supported in northern temperate altricial birds. In contrast, the effects of weather during the breeding season often affects the

---

E-mail address: bernt-erik.sather@chembio.ntnu.no (B.-E. Sæther)

ADVANCES IN ECOLOGICAL RESEARCH, VOL. 35
0065-2504/04 $35.00 DOI 10.1016/S0065-2504(04)35009-9

population fluctuations of many nidifugous species and species living under arid conditions, thus supporting the tap-hypothesis. However, the dynamical consequences of these interspecific differences in the timing of the population limitation cannot be properly understood without modelling and estimating the effects of the density dependence as well as changes in the mean and variance of the relevant climate variables.

II. INTRODUCTION

It is well accepted that changes in climate may result in changes in the population size of birds. In this chapter, we will consider the exact mechanisms by which this may occur. Understanding the mechanisms is essential if we wish to make predictions about the ecological consequences of the expected changes.

Population size depends upon birth rates and death rates and how these interact with density. The variation in these relationships is also important. By understanding how these interact to affect variation in population size, we can examine how climate change may affect the future fluctuations of bird populations. There are three major issues to consider in understanding how climate change may result in changes in population size.

(1) Are the climate responses density dependent or density independent? For example, changes in rainfall could kill an equal proportion of adults or nestlings regardless of density and so result in a density independent change in these parameters. Alternatively, changes in rainfall might change the availability of food or nesting sites and so alter the relationship between density and either survival or breeding success. Turchin (1995) provides a review of these issues.

(2) Does climate change largely affect the reproductive success or the survival? We review studies of climate effects on fluctuations of bird populations to examine evidence for two hypotheses for under which conditions climate is likely to most strongly affect avian population dynamics. According to the tub-hypothesis, originating back to Lack (1954), we would expect fluctuations in population size to be closely related to climate variation during the non-breeding season because in combination with density dependence, the weather conditions determine the number of birds surviving during this critical period of the year. In contrast, according to the tap-hypothesis, we would expect the change in population size from one year to another to be related to the weather during the breeding season because this will influence the inflow of new recruits into the population the following year.

(3) Does climate change primarily affect the mean values or the variability? As examples of the consequences for mean values, climate change may

result in changes in food supply that shifts the relationship between survival and density (for example, food is scarcer so starvation occurs at lower population sizes). Alternatively, warmer weather may consistently change the breeding success. As examples of how variability may be affected, climate change may alter the frequency of severe cold weather or droughts and change the probability of periods of high mortality. As another example, changes in the flooding regime may affect the likelihood of catastrophic floods or drought in the breeding season. Of course, mean and variance often are interrelated and we will present an approach to model the effects on the population dynamics of changes of mean and variances in climate variables.

The purpose of this chapter is to review studies that have analysed the effects of local weather and large-scale climate phenomenon on fluctuations in the size of bird populations. We then outline a general theoretical framework based on Lande *et al.* (2003) for analysing avian population dynamics and describe how an expected change in climate may affect avian population dynamics and thus help resolve the three issues described above.

III. IMPACT OF WEATHER ON AVIAN POPULATION DYNAMICS

As shown in Table 1 we have reviewed all studies we were able to find that related the effects of local weather on avian population dynamics. This shows that weather conditions during both the breeding season and the non-breeding season may have significant effects on the fluctuations in the size of bird populations.

Unfortunately, this table cannot be used to quantitatively examine whether the weather primarily affects the population dynamics through an effect on recruitment (the tap-hypothesis) or on the number of individuals that survive the non-breeding season (the tub-hypothesis). The main reason for this is that few studies have examined the relative contributions of different climate variables collected at different times of the year to temporal variation in the fluctuations in population size. Few studies have also accounted for the effects of density dependence.

However, some general patterns seem to emerge from the studies assembled in Table 1:

(1) Climate conditions during the non-breeding season often affect changes in the size of populations of northern temperate altricial birds.

Table 1 Studies examining the effects of local climate on annual variation in population size N, change in population size from one year to the next ΔN, residual variation after accounting for the effects of population size or on some population index

Species	Locality	Dependent variable	Independent variable[a]	Source
Anseriformes				
Eurasian wigeon *Anas penelope*	Lake Myvatn, Iceland	ΔN (males)	Date of water becoming ice free*	Gardarsson and Einarsson (1997)
Mute swan *Cygnus olor*	Copenhagen, Denmark	N	Number of days with ice-cover of water	Bacon and Andersen-Harild (1989)
Passeriformes				
Barn swallow *Hirundo rustica*	Kraghede, Denmark	ΔN	Precipitation March, South Africa	Møller (1989)
Bewick's wren *Thryomanes bewickii*	California, USA	Population index	Winter temperature	Verner and Purcell (1999)
Blue tit *Parus caeruleus*	Wytham Wood, UK	ΔN	Early spring temperature	Slagsvold (1975)
Blue tit	Oranje Nassu's Oord, Netherlands	ΔN	Temperature Jan–March***	Slagsvold (1975)
British resident farmland birds	Common bird censuses, UK	ΔN	Number of snowdays, precipitation in Mar–Apr	Greenwood and Baillie (1991)
British resident woodland birds	Common bird censuses, UK	ΔN	Number of snowdays, precipitation in Mar–Apr	Greenwood and Baillie (1991)
Darwin's medium ground finch *Geospiza fortis*	Galapagos Islands	N	Rainfall	Grant *et al.* (2000)
Darwin's cactus ground finch *Geospiza scandens*	Galapagos Islands	N	Rainfall	Grant *et al.* (2000)

Species	Location	Response	Climate variable	Reference
Dipper *Cinclus cinclus*	Lyngavassdraget, southern Norway	ΔN	Number of days with ice-cover of water. Mean winter temperature	Sæther et al. (2000a)
Galapagos mockingbird *Nesomimus parvulus*	Galapagos Islands	N	Rainfall	Curry and Grant (1989)
Garden warbler *Sylvia borin*	Common bird censuses, UK	Population index	Sahel rainfall[b]	Baillie and Peach (1992)
Grassland birds	Great plains	N	Summer precipitation and temperature	George et al. (1992)
Great tit *Parus major*	Forest of Dean, UK	ΔN	Early spring temperature	Slagsvold (1975)
Great tit	Wytham Wood, UK	ΔN	Early spring temperature	Slagsvold (1975)
Great tit	Lemsjøholm, Finland	ΔN	Temperature Feb**	Slagsvold (1975)
Great tit	Wytham Wood, UK	Residual variation	Winter–spring temperature	Lebreton (1990)
Great tit	Farmland habitat, UK	Change in population index	Nov temperature	O'Connor (1980)
Great tit	Woodland habitat	Change in population index	Nov temperature	O'Connor (1980)
Great tit	Lemsjøholm, Finland	N	Dec–Feb temperature	von Haartman (1973)
Great tit	Oulu, Finland	ΔN	Apr temperature*	Orell (1989)
Grey-headed junco *Junco hyemalis*	Central Arizona, USA	N	Precipitation May–Jun**	Martin (2001)
House wren *Troglodytes aedon*	California, USA	Population index	4 years running mean of annual precipitation** Lowest temperature the four preceding months before breeding**	Verner and Purcell (1999)

(Continued)

Table 1 Continued

Species	Locality	Dependent variable	Independent variable[a]	Source
Orange-crowned warbler *Vermivora celata*	Central Arizona, USA	N	Precipitation May–Jun***	Martin (2001)
Red-faced warbler *Cardellina rubifrons*	Central Arizona, USA	N	Precipitation May–Jun	Martin (2001)
Sand martin *Riparia riparia*	Maglarp, Sweden	N	Summer temperature$_{t-1}$*	Persson (1987)
Sand martin	River Tisza, Hungary	N	African rainfall	Szép (1995)
Sedge warbler *Acrocephalus schoenobaenus*	Common bird censuses, UK	Population index	Sahel rainfall[a]	Baillie and Peach (1992)
Sedge warbler	Extensive population censuses, NL	Population index	Sahel rainfall***	Foppen *et al.* (1999)
Song sparrow *Melospiza melodia*	Manadarte Island, Canada	N	Winter temperature	Arcese *et al.* (1992)
Trans-Saharian migrants (14 spp.)	Europe	ΔN	African rainfall	Marchant (1992)
Virginia warbler *Vermivora virginiae*	Central Arizona, USA	N	Precipitation May–Jun***	Martin (2001)
Strigiformes Northern spotted owl *Strix occidentalis caurina*		Recruitment	Precipitation and temperature in early nesting period. Winter precipitation	Franklin *et al.* (2000)
Gruiformes Coot *Fulica atra*	Aalsmeer, Netherlands	N_{t+1}/N_t	Number of ice-days	Cavé and Visser (1985)

Cicconiiformes				
Adélie penguin *Pygoscelis adeliae*	Ross Island, Antarctica	N_{t+1}/N_t	Monthly maximum ice extent$_{t-5}$	Wilson *et al.* (2001)
Black-crowned night heron *Nycticorax nycticorax*	Southern France	N^{**}	Rainfall Sahara	Den Held (1981)
Common buzzard *Buteo buteo*	Niederrhein, Germany	N	Winter temperature, snow cover	Kostrzewa and Kostrzewa (1991)
Golden plover *Pluvialis apricaria*	Peak District, UK	ΔN	Winter temperature*	Yalden and Pearce-Higgins (1997)
Goshawk *Accipiter gentilis*	Niederrhein, Germany	N	Winter temperature, snow cover	Kostrzewa and Kostrzewa (1991)
Goshawk	Eastern Westphalia, Germany	ΔN	Composite weather variable for the breeding season–summer	Krüger and Lindström (2001)
Grey heron *Ardea cinerea*	England and Wales	N	Winter temperature	Lack (1966), Reynolds (1979)
Kestrel *Falco tinnunculus*	Niederrhein, Germany	N	Winter temperature**, snow cover*	Kostrzewa and Kostrzewa (1991)
Prairie falcon *Falco mexicanus*	Idaho, USA	N	Winter drought	Steenhof *et al.* (1999)
Purple heron *Ardea purpurea*	Nieuwkoop, Netherlands	N^{**}	Rainfall Sahara	Den Held (1981), Cavé (1983)
Squacco heron *Ardeola ralloides*	Southern France	$N(*)$	Rainfall Sahara	Den Held (1981)
Columbiformes				
Zanaida dove *Zenaida aurita*	Cayo del Agua, Puerto Rico	N	Rainfall during the first year***	Rivera-Milán and Schaffner (2002)

(Continued)

191

Table 1 Continued

Species	Locality	Dependent variable	Independent variable[a]	Source
Galliformes				
Black grouse *Tetrao tetrix*	Glen Tanar, UK	Recruitment	Rainfall summer	Moss (1986)
California quail *Callipepla californica*	California, USA	N	Dec–Apr precipitation	Botsford et al. (1988)[a]
Capercaillie *Tetrao urogallus*	Glen Tanar, UK	Recruitment	Rainfall June[c] April temperature[d]	Moss and Oswald (1985), Moss et al. (2001)
Capercaillie	Vegårdshei, southern Norway	Population index	Rainfall June–July, date of snowmelt	Slagsvold and Grasaas (1979)
Eastern wild turkey *Meleagris gallopavo*	Wisconsin, USA	N	Temperature Mar–Apr	Rolley et al. (1998)
Grey partridge *Perdix perdix*	Washington, USA	Residual variation of change in population size	Winter temperature, winter precipitation	Rotella et al. (1996)
Northern bobwhite *Colinus virginianus*	Texas, USA	Population index	Summer temperature, fall precipitation	Lusk et al. (2002)
Rock ptarmigan *Lagopus mutus*	Cairngorms, UK	Log N	June temperature[e]*	Watson et al. (2000)

[a] Statistics refer to tests of significance of annual variation. $*p < 0.05$, $**p < 0.01$, $***p < 0.001$. Brackets denote marginal significant effects.

[b] Significant correlation ($p < 0.05$) between overwinter losses and Sahel rainfall.

[c] Breeding success inversely related to the number of days with rain during the first 10 days of June.

[d] Brood size declined with April temperature.

[e] Brood size increased with June precipitation (Watson et al., 1998).

This operates mainly through a weather-dependent loss of birds (see Møller, 1989; Baillie and Peach, 1992). This provides support for the tub-hypothesis.

(2) Climate conditions during the breeding season may affect population fluctuations of Galliformes species through an effect on recruitment rate. This supports the suggestion of Sæther et al. (1996) that the population dynamics of nidifugous and altricial birds differ in many important aspects.

(3) Population fluctuations of species living under arid conditions seem to be strongly influenced by the weather conditions during the breeding season (e.g., Grant et al., 2000; Martin, 2001) because successful breeding may be possible only under favourable weather conditions.

These general patterns are also supported by some especially detailed analyses of the effects of weather on avian population dynamics. Greenwood and Baillie (1991) studied the factors affecting fluctuations of British bird populations, based on the Common Bird Censuses, in relation to weather in different periods of the year and showed that winter temperature was an important significant explanatory variable for the change in population size of farmland species. Nine species showed significant correlations between population change and temperature recorded during the period November–April. In contrast, only one such significant correlation was recorded for the period May–October. Similarly, in a very detailed examination of the relationship between fluctuations in the size of the tit populations in Wytham Wood outside Oxford in England and weather at different times of the year, Slagsvold (1975) found the highest significant correlation coefficients for periods between the end of winter and early spring. Similarly, in a Finnish population of great tit Parus major annual variation in the change in population size was correlated to mid-winter temperatures (Table 1).

As expected from the correlations between different weather variables and avian population dynamics (Table 1), large-scale climate phenomena such as the NAO or the El Niño Southern Oscillation (ENSO) that influence local climate over large areas (Hurrell, 1995; Hurrell et al., 2003), often explain a significant proportion of annual variation in the size of bird populations (Table 2). Such a relationship is found in long-lived species such as the emperor penguin Aptenodytes forsteri (Barbraud and Weimerskirch, 2001) or blue petrel Halobaena caerulea (Barbraud and Weimerskirch, 2003) in Antarctica as well as in small passerines such as the dipper Cinclus cinclus (Sæther et al., 2000a; Loison et al., 2002). In both these cases, the large-scale climate phenomena mainly operate through an effect on the losses during the non-breeding part of the season.

Several of the studies included in Table 2 including both terrestrial and marine species involve analyses of the demographic consequences of

Table 2 Studies examining the effects of large-scale climate phenomena on annual variation in population size N, change in population size from one year to the next ΔN, residual variation after accounting for the effects of population size or on some population index

Species	Locality	Dependent variable	Independent variable	Source
Passeriformes				
Collared flycatcher *Ficedula alibollis*	Dlouha Loucka, Czech Republic	$\ln(N)$	$NAO_w^{(*)}$	Sætre et al. (1999)
Darwin's medium ground finch	Galapagos Islands	N	El Niño occurrences	Grant et al. (2000)
Darwin's cactus ground finch	Galapagos Islands	N	El Niño occurrences	Grant et al. (2000)
Dipper	Lyngavassdraget, southern Norway	Residual variation	NAO_w^{**}	Sætre et al. (2000a)
Farmland birds (15 species)	Scotland	Mean population size	NHA_{t-1}	Benton et al. (2002)
Galapagos mockingbird	Galapagos Islands	N	El Niño occurrences	Curry and Grant (1989)
Great tit	Europe	Residual variation	NAO_w^{a}	Sætre et al. (2003)
House martin *Delichon urbica*	Bonn, Germany	Residual variation	NAO_w^{*}	Stokke et al. (2004)
Pied flycatcher *Ficedula hypoleuca*	Dlouha Loucka, Czech Republic	$\ln(N)$	NAO_w	Sætre et al. (1999)
Pied flycatcher	Europe	Residual variation	NAO_w^{b}	Sætre et al. (2003)
Ciconiiformes				
Adélie penguin	Ross Island, Antarctica	N_{t+1}/N_t	SOI^{**}	Wilson et al. (2001)
Blue petrel	Kerguelen Islands, southern Indean Ocean	N	SOI_w	Barbraud and Weimerskirch (2003)

		Residual variation		
Continental great cormorant *Phalacrocorax carbo sinensis*	Coast of Central Europe		NAO_w [c, d]	S. Engen et al. (in press)
Galapagos penguin *Spheniscus mendiculus*	Galapagos islands	N	El Niño/La Niña-occurrences	Dee Boersma (1998)
King penguin *Aptenodytes forsteri*	Terre Adélie, Antarctica	ΔN	SST	Barbraud and Weimerskirch (2001)
Northern fulmar *Fulmarus glacialis*	Eynhallow, UK	N	NAO_w**	Thompson and Ollason (2001)
Snow petrel *Pagodroma nivea*	Adélie Land, Antarctica	N	El Niño-occurrences	Chastel et al. (1993)
Others				
Different seabird species	Christmas Island, Pacific Ocean	N	El Niño occurrence	Schreiber and Schreiber (1984)

*$p < 0.05$, **$p < 0.01$. Brackets denote marginal significance.
NAO_w denotes the Northern Atlantic Oscillation Index during winter, SOI is the Southern Oscillation Index, SST is the annual average sea surface temperatures, NHA is the Northern Hemisphere Temperature Anomalies (Jones et al., 1999).
[a] Significant contribution to the residual variation in 8 of 40 populations.
[b] Significant contribution to the residual variation in 6 of 29 populations.
[c] Significant contribution to the residual variation in 5 of 22 colonies.
[d] Regression coefficient of residual variation of change in population size on NAO was positive in 16 of 22 colonies.

El Niño/La Nina events. The most detailed understanding has probably been obtained for the effects of El Niño years on the population dynamics of several species living on the Galapagos Islands (Grant, 1986; Curry and Grant, 1989; Grant and Grant, 1989; Grant *et al.*, 2000). These studies have demonstrated that the influence of El Niño-related rainfall is the major driving force for many of the demographic and evolutionary changes that have been recorded in many of those populations over the past decades. For instance, in years of abundant rainfall up to eight breeding attempts could be made by an individual pair of Darwin's medium ground finch *Geospiza fortis* or cactus finch *G. scandens*, whereas no breeding occurred at all in years with drought (Grant and Grant, 1992).

An examination of Table 2 shows that most studies of the effects of large-scale climate pattern on avian population dynamics consist of one study of a single species. The only exception is a study by Sæther *et al.* (2003) of two European hole-nesting species (pied flycatcher *Ficedula hypoleuca* and great tit) and another study of continental great cormorant *Phalacrocorax carbo sinensis* (S. Engen *et al.*, in press). A comparison of the relationship between NAO and changes in population size of great tit and pied flycatcher across Europe showed latitudinal gradients in the contribution of a large-scale climate phenomenon to the population fluctuations of those two species (Sæther *et al.*, 2003). In both species, there was an increase in the absolute value of the regression coefficient of changes in population size on NAO, β_{NAO}, with latitude (correlation coefficient = 0.463, $p = 0.003$, $n = 40$ for great tit and correlation coefficient = 0.378, $p = 0.048$, $n = 29$ for pied flycatcher). However, the sign of the effects of NAO differed in both species over relatively short distances: in some populations, a positive relationship was present, whereas in others population size decreased with increasing NAO. This suggests that microgeographic variations relating large-scale climate phenomena such as NAO are related to local weather and may generate differences in local population dynamics across short distances (see also Mysterud *et al.*, 2000). However, this latitudinal gradient in the effects of NAO affected the population dynamics of the two species differentially. In the great tit, the proportion of the population variability explained by NAO increased with latitude. In contrast, in the pied flycatcher such an increase with latitude was not found because the residual component of the environmental variance also increased with latitude, resulting in a latitudinal increase in the total environmental stochasticity. This clearly illustrates that the effects of climate on bird population dynamics may differ geographically (see also Marchant, 1992). This is also expected from the large regional variation that often has been recorded in the relationship between climate and different demographic traits (e.g., Sanz, 2002, 2003; Sanz *et al.*, 2003; Visser *et al.*, 2003).

IV. MODELLING POPULATION DYNAMICAL CONSEQUENCES OF CHANGES IN CLIMATE

Substantial evidence suggests that the global climate now changes at a rate that has hardly been experienced in the past history of the earth (Parker *et al.*, 1994), even though the mechanisms are not yet fully understood (Tett *et al.*, 1999). These changes are also expected to continue in the near future (Houghton *et al.*, 2001). Basically, the effects on the population dynamics can occur in three different ways: (1) A change can occur in the variance of the stochastic weather variable. (2) Demographic variables may change in a non-linear way with a climate variable (Visser *et al.*, 2003; Visser *et al.*, 2004, this volume). (3) A temporal trend may occur in the climate variable. Accordingly, a continued increase in mean temperature is expected in the future (Stott and Kettleborough, 2002). Here we will illustrate how such changes can be included into simple population dynamical models, following the general approach of Lande *et al.* (2003).

A. Changes in Mean Demographic Variables

If climate results in a consistent density independent change in the birth rate or death rate, then it is conceptually very straightforward to predict the population response. The magnitude of the response will depend upon the strength of density dependence. Thus, if the regulation is weak the population will be affected more than if the regulation is strong. In many cases, climate change will result in changes in the nature of density dependence. A straightforward example is where climate change alters the area of suitable habitat (for example, by sea level rise altering the area of habitat or climate change affecting the area of a given vegetation type). Sutherland (1996) showed for a migratory population that the ratio of population change to habitat loss in, say, the wintering grounds was directly proportional to the ratio of winter density dependence to total density dependence. The equivalent result for the loss of breeding habitat is the ratio of density dependence in the breeding grounds to total density dependence.

In practice, it can be difficult to measure density dependence. Liley (1999) made detailed behavioural observations of the territorial preferences of ringed plovers *Charadrius hiaticula* and determined the factors that affected breeding output and used a game theoretical model to predict the relationship between breeding output and density. He then used this model to predict the consequences of sea level rise, which affects beach width, which was important in determining breeding success. Furthermore, human disturbance was shown to be important and the effects of changes in this were predicted. Climate change is likely to result in changes in tourism.

There are difficulties in distinguishing mean changes from changes in stochasticity and this is examined in the following section.

B. Changes in the Environmental Stochasticity

To examine how a change in the variance in a climate variable affects the long-term population dynamics, consider a simple model describing density independent growth in a random environment $N(t + 1) = \lambda(t)N(t)$ (Lewontin and Cohen, 1969). Here $N(t)$ is population size in year t, and $\lambda(t)$ is the population growth rate in year t, which is assumed to have the same probability distribution each year with mean $\bar{\lambda}$ and variance σ_λ^2. If the population size is sufficiently large to ignore demographic stochasticity, the environmental stochasticity is $\sigma_e^2 = \sigma_\lambda^2 = \mathrm{var}((\Delta N/N)|N)$, where ΔN is the change in population size between t and $t + 1$. If $X(t) = \ln N(t)$, we get on the natural log scale $X(t + 1) = X(t) + r(t)$, where $r(t) = \ln \lambda(t)$. The growth rate during a period of time t, $(X(t) - X(0))/t$, has mean $s = Er$ and variance $\mathrm{var}(r)/t = \sigma_r^2/t$ (E denotes the expectation). The stochastic population growth rate s is the constant mean slope that the trajectories approach when plotted on log scale at long time intervals. This leads to the approximation

$$s \approx \ln \bar{\lambda} - \frac{\sigma_r^2}{2} \quad \text{and} \quad \sigma_r^2 \approx \frac{\sigma_\lambda^2}{\bar{\lambda}^2} = \frac{\sigma_e^2}{\bar{\lambda}^2} \tag{1}$$

for the mean and the variance, respectively (Lande et al., 2003). Simulations of this model show that after some time, all sample paths are below the trajectory for the deterministic model (Figure 1). Stochastic effects reduce the mean growth rate of a population on a logarithmic scale by approximately $\sigma_r^2/2$, compared to that in a constant environment. Thus, an increase in the stochasticity σ_r^2 will decrease the long-term growth rate.

So far, we have only considered the effects of adding stochasticity while keeping the deterministic growth rate $r = \ln(\bar{\lambda})$ constant. This is equivalent to keeping the expected fecundity and mortality constant. However, changes in climate may affect these expectations, and hence r, as is schematically presented in Figure 1. In this case, the effects on long-term population growth rate may be different from only changing the environmental stochasticity. For instance, in stock-recruitment models that are commonly used in fisheries (Quinn and Deriso, 1999) stochasticity is usually added to the exponent

$$R = g(S)e^{\sigma\varepsilon}, \tag{2}$$

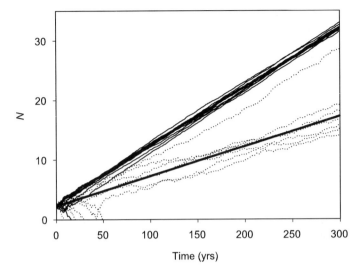

Figure 1   Simulating 10 sample paths of density-independent growth under the influence of different levels of environmental stochasticity $\sigma_e^2 = 0.001$ (dotted lines) and $\sigma_e^2 = 0.01$ (solid lines). The thick lines show the expected long-run growth of the populations. Other parameters were $r = 0.01$ and $\sigma_d^2 = 0.5$.

where R and S are recruitment and stock, respectively, $g(\cdot)$ is a function of population density and ε is a standardised variable with zero mean and unit variance. Assuming that ε is normally distributed, we see that the expected recruitment is

$$E(R|S) = g(S)e^{\sigma^2/2} > g(S).$$

In this case, stochasticity increases the expected recruitment because the stochasticity is modelled in a way that increases the deterministic growth rate r. If we instead would like to evaluate the stochastic effects alone as in Section III, we should rather use a stock recruitment model of the form $R = g(S)W$, where W is a stochastic variable with expectation 1. In the model in Eq. (2), this could be done by choosing $W = \exp(\sigma\varepsilon - (1/2)\sigma^2)$. Then the expected recruitment is kept constant and the stochastic growth rate will decrease with increasing environmental stochasticity.

These examples illustrate that great care must be taken in considering how environmental stochasticity enters the model when the effects on population dynamics of changes in a climate variable is modelled.

C. Trends in Climate

Over larger part of the globe, the climate is changing and is expected to do so also in the future (Houghton et al., 2001). For instance, an increase in temperature has occurred over large areas the last decades (Mann et al., 1998), which has affected the demography of bird populations (e.g., Sanz, 2003; Visser et al., 2003). A proper modelling of the population dynamical consequences of such trends in climate requires separation of the effects of density-dependence from the stochastic influences on the population dynamics (Lebreton, 1990; Rotella et al., 1996; Dennis and Otten, 2000; Sæther et al., 2000a). We have elsewhere suggested (Sæther et al., 2000b, 2002a,b; Sæther and Engen, 2002) that the theta-logistic model of density regulation where the change in population size from one year to the next can be written in the form

$$\Delta N = r_1(t)N - \bar{r}N\frac{N^\theta - 1}{K^\theta - 1} \tag{3}$$

has many useful properties. Here K is the carrying capacity defined as the population size where the expected change in N is zero, $r_1(t)$ is the population growth at population size $N = 1$ that fluctuates with mean $\bar{r}$ and variance $\sigma_r^2 = \sigma_e^2 + \sigma_d^2/N$, where σ_d^2 is the demographic variance, i.e., random variation in individuals in their contribution to fitness (Lande et al., 2003). The parameter θ describes the form of the density regulation (Sæther et al., 2002a). When θ approaches zero, we get the well-known loglinear model often used in ecological modelling (Royama, 1992). For large $\theta (\theta \gg 1)$, density regulation first starts to act for population sizes closer to K. When $\theta = 1$, we get the logistic model.

The model for $r_1(t)$ may alternatively be written as $r_1(t) = \bar{r} + \sigma_e U + \sigma_d V/\sqrt{N}$, where U and V are independent stochastic variables. We now include in the model a climate variable, Z, with a trend (e.g., spring temperature in northern Europe), writing

$$r_1(t) = \bar{r} + \sigma_e U + \sigma_d V/\sqrt{N} + \beta Z, \tag{4}$$

where β is the regression coefficient for the effect of Z. Inserting Eq. (4) into the equation for the theta-logistic model (Eq. (3)) gives, after some algebra, the expected change in N (conditioned on N)

$$E(\Delta N|N) = (\bar{r} + \beta EZ)N \left(1 - \frac{N^\theta - 1}{K_\beta^\theta - 1}\right) \tag{5}$$

where $K_\beta = [1 + (K^\theta - 1)(\bar{r} + \beta EZ)/\bar{r}]^{1/\theta}$ for $\theta \neq 0$ and $K_\beta = K^{(\bar{r} + \beta EZ)/\bar{r}}$ for $\theta = 0$.

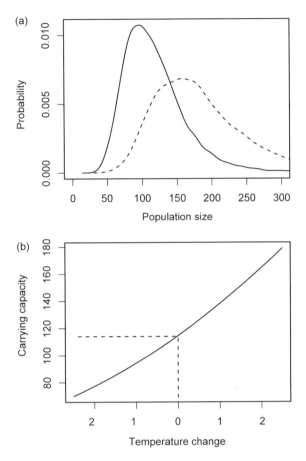

Figure 2   The distribution of the population size (a) before (full lines) and after an increase in winter temperature of 2.5°C (stippled lines). (b) The change in carrying capacity K as a function of a change in mean winter temperature (°C). The dotted line indicates the current situation (from Sæther *et al.*, 2000a).

We then want to examine how the variance in Z contributes to the variation in N. Writing σ_Z^2 for the variance in Z through time we have

$$\text{var}(\Delta N | N) = (\sigma_e^2 + \beta^2 \sigma_Z^2) N^2 + \sigma_d^2 N. \tag{6}$$

This shows that temporal stochastic fluctuations in the covariate Z add a term $\beta^2 \sigma_Z^2$ to the environmental variance. By these modifications of the parameters we can explore effects of changes in the mean and the

Table 3 Summary of main effects of climate on avian population dynamics. According to the tub-hypothesis, fluctuations in population size will be closely related to climate variation during the non-breeding season because in combination with density dependence the weather conditions determine the number of birds surviving during this critical period of the year. According to the tap-hypothesis, annual variation in population size is related to weather during the breeding season through the inflow of new recruits into the population the following year

Group of species	Main effect
Temperate altricial birds	Climate affects in combination with density dependence the losses during the non-breeding season (the tub-hypothesis)
Temperate nidifugous species	Climate-induced changes in recruitment affect changes in population size (the tap-hypothesis)
Species in arid environments (with larger clutch sizes)	Climate-induced changes in recruitment affect changes in population size (the tap-hypothesis)
Long-lived species	Variation in climate introduce time-lags in the population dynamics

variance in a climate variable on the expected equilibrium population size (Eq. (5)) or the relative contribution of stochastic variation in Z to the variability in population size (Eq. (6)).

The above derivation is based on σ_Z^2 being a white noise process with no correlation among values in different years. However, we know that many climate variables are now showing clear trends (Houghton *et al.*, 2001). Our modelling approach can be extended to model the consequences of such climatic trends. If the process is autocorrelated with $\text{corr}(Z_t, Z_{t+h}) = \rho_h$, the process can be approximated (Lande *et al.*, 1995) by a white noise process with σ_Z^2 replaced with $\left(1 + 2\sum_{h=1}^{\infty} \rho_h\right)\sigma_Z^2$ and substituted in Eq. (6).

This modelling approach was utilised by Sæther *et al.* (2000a) in their analysis of changes in climate will affect the dynamics of a population of dipper in southern Norway. Assuming a loglinear model of density regulation ($\theta = 0$ in Eq. (3)), environmental variance from winter temperature contributed about 50% to the total environmental variance. This occurred because a high proportion of juveniles as well as adults died during cold winters (Loison *et al.*, 2002). Accordingly, there was a positive relationship between change in population size and mean winter temperature, number of days in which the water was ice-covered and winter-NAO (Sæther *et al.*, 2000a). As a consequence, an increase in winter climate would change the stationary distribution of population sizes (Figure 2a) and result in a non-linear increase in the expected equilibrium population size (see Eq. (5)) with mean winter temperature (Figure 2b).

V. DISCUSSION

In this chapter, we show that local weather (Table 1) as well as regional climate pattern (Table 2) may strongly influence avian population dynamics. Unfortunately, the number of studies is still insufficient for a quantitative examination of the tub- and tap-hypotheses presented in Section I. However, some patterns seem to be present (Table 3). For instance, a majority of studies found that climate outside the breeding season affected fluctuations in population numbers, in many cases operating on variation in survival rates (Cavé, 1983; Møller, 1989; Baillie and Peach, 1992; Szép, 1995; Barbraud and Weimerskirch, 2001; Loison et al., 2002; Stokke et al., 2004). This strongly supports the tub-hypothesis of Lack (1954) that population limitation often occurs during the non-breeding season.

Weather in the breeding season also affects the population fluctuations of many species. Evidence suggests that weather conditions just prior to or during the breeding season influence annual variation in change in population size of many galliform species (Table 1) through an effect on the recruitment thus supports the tap-hypothesis. Accordingly, this suggests the limitation of populations of nidifugous and altricial species occurs at different times of the year. Although we realize that key-factor analyses (Podoler and Rogers, 1975) are based on several assumptions that are not justified (independent data points, no age structure and no sampling error), we believe that the different patterns recorded in key factors analyses of these two groups of birds also support this conclusion (Figure 3). The major k-factor that explains the highest proportion of the changes in population size is recorded during the non-breeding season in 93% of the studies involving altricial birds. In contrast, the corresponding figure was 7% for nidifugous birds. Thus, we hypothesize (Table 3) that climate during the breeding season will have a stronger influence on the population fluctuations of nidifugous than of altricial species that will be more strongly influenced by the weather during the non-breeding season.

However, a strong effect on the breeding season of weather during the breeding season is also found in many passerine species that live in arid environments (Table 1). This may suggest (Table 3) that the tap-hypothesis also applies to the population dynamics of these species if the critical mean resource supply for breeding is below a certain level.

The studies reviewed here (Tables 1 and 2) reveal that we are far from a general understanding of how climate changes will affect avian population dynamics. As is seen in Section IV, this requires modelling of the density-dependent effects (e.g., Lebreton, 1990; Rotella et al., 1996; Yalden and Pearce-Higgins, 1997; Sætre et al., 1999; Watson et al., 2000; Krüger and Lindström, 2001; Sæther et al., 2002a). Furthermore, demographic stochasticity

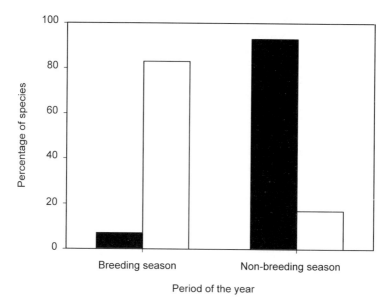

Figure 3 Seasonal variation in the principal *k*-factor, i.e., the key factor (Podoler and Rogers, 1975) explaining the highest proportion of the annual variation in population change in altricial (black columns) (*n* = 14 species) and in nidifugous (white columns) (*n* = 6 species) birds. Additional sources to those listed in Sæther *et al.* (1996) are Potts (1986), Stenning *et al.* (1988), Møller (1989) and Baillie and Peach (1992).

also influences the dynamics of populations, particularly those that are small (see Sections IV.B and C). However, long-term data necessary for estimating those parameters are likely to be available only for a limited set of species. Fortunately, some recent comparative analyses (Sæther and Engen, 2002; Sæther *et al.*, 2002a; Sæther *et al.*, in press-a,b) indicate some of these parameters can be approximated from knowledge of basic life history variables. This will facilitate a comparative approach to quantitative analysis of how different types of species will respond to different climate change scenarios.

That such general patterns may be present are indicated by the analyses of Pimm (1991) of the population responses of different British birds to the especially cold winter of 1962/63. Based on an initial analysis by O'Connor (1981), he demonstrated that the rate of recovery of the populations after this winter increased with the species' potential egg-production per breeding season. Furthermore, small-bodied species declined more during a later harsh winter than larger-bodied species. This suggests that population responses to climate changes may be predictable from basic ecological traits. Deriving such predictions will, however, require parameterisation of stochastic population models such as those presented in Section IV.

The modelling framework presented in Section IV must be further extended to include important aspects of how climate is expected to affect avian population dynamics. For instance, a general feature of many bird species is that they show synchronous population fluctuations over large areas (e.g., Ranta *et al.*, 1998; Paradis *et al.*, 2000 but see also Ringsby *et al.*, 2002). Theoretical analyses have shown that spatial correlations in the environment, often induced by a common climate, may affect the spatial scaling of population fluctuations (e.g., Lande *et al.*, 1999). Because synchrony in population fluctuations affect the risk of extinction (Engen *et al.*, 2002), we must understand how large-scale changes in climate will affect the spatial scaling of fluctuations in the size of bird populations. Some evidence does indeed suggest that the consequences of climate changes will depend on the landscape characteristics such as the degree of habitat fragmentation (Foppen *et al.*, 1999; Franklin *et al.*, 2000; Ringsby *et al.*, 2002). Migrating birds will represent a special challenge for bird ecologists in this respect (e.g., Sillett *et al.*, 2000). Furthermore, lagged responses in the effects of climate on the population dynamics or demography often are apparent in populations of long-lived species (Barbraud and Weimerskirch, 2001, 2003; Thompson and Ollason, 2001). We, therefore, need to extend the models in Section IV.C to include age (Lebreton and Clobert, 1991; Lande *et al.*, 2002) to distinguish changes in population size due to climate variation from population fluctuations due to delayed age-structure effects. Another aspect that is poorly understood is how climate induced selection pressures will affect individual characteristics (e.g., Møller, 2002; Møller and Erritzøe, 2003) or demographic variables (see review in Dunn, 2004) and how such changes will influence the population dynamical characteristics. Finally, the understanding of how climate variation affects trophic interactions and interspecific relationships (but see Stenseth *et al.*, 2002) are almost non-existent in birds although studies of long-lived seabirds (Tables 1 and 2, Thompson and Ollason, 2001) strongly suggest that such interactions are important.

ACKNOWLEDGEMENTS

This study was financed by the Research Council of Norway (Klimaeffekt-programmet). We are grateful to Anders Pape Møller and Jean Clobert for comments on the manuscript.

REFERENCES

Arcese, P., Smith, J.N.M., Hochachka, W., Rogers, C.M. and Ludwig, D. (1992) *Ecology* **73**, 805–822.

Bacon, P.J. and Andersen-Harild, P. (1989) In: *Lifetime Reproduction in Birds* (Ed. by I. Newton), pp. 363–386. Academic Press, London.

Baillie, S.R. and Peach, W.J. (1992) *Ibis* **134(suppl. 1)**, 120–134.

Barbraud, C. and Weimerskirch, H. (2001) *Nature* **411**, 183–186.

Barbraud, C. and Weimerskirch, H. (2003) *Proc. R. Soc. Lond. B* **270**, 2111–2116.

Benton, T.G., Bryant, D.M., Cole, L. and Crick, H.Q.P. (2002) *J. Appl. Ecol.* **39**, 673–687.

Botsford, L.W., Wainwright, T.C., Smith, J.T., Mastrup, S. and Lott, D.F. (1988) *J. Wildl. Manage.* **52**, 469–477.

Cavé, A.J. (1983) *Ardea* **71**, 217–224.

Cavé, A.J. and Visser, J. (1985) *Ardea* **73**, 129–138.

Chastel, O., Weimerskirch, H. and Jouventin, P. (1993) *Oecologia* **94**, 278–285.

Curry, R.L. and Grant, P.R. (1989) *J. Anim. Ecol.* **58**, 441–463.

Dee Boersma, P. (1998) *Condor* **100**, 245–253.

Den Held, J.J. (1981) *Ardea* **69**, 185–191.

Dennis, B. and Otten, M.R. (2000) *J. Wildl. Manage.* **64**, 388–400.

Dunn, P. (2004) *Adv. Ecol. Res.*

Engen, S., Lande, R. and Sæther, B.-E. (2002) *Am. Nat.* **160**, 439–451.

Engen, S., Lande, R., Sæther, B.-E. and Bregnballe, T. *J. Anim. Ecol.* (in press).

Foppen, R., Ter Braak, C.J.F., Verboom, J. and Reijnen, R. (1999) *Ardea* **87**, 113–127.

Franklin, A.B., Anderson, D.R., Gutiérrez, R.J. and Burnham, K.P. (2000) *Ecol. Monogr.* **70**, 539–590.

Gardarsson, A. and Einarsson, A. (1997) *J. Anim. Ecol.* **66**, 439–451.

George, T.L., Fowler, A.C., Knight, R.L. and McEwen, L.C. (1992) *Ecol. Appl.* **2**, 274–284.

Grant, P.R. (1986) *Ecology and Evolution of Darwin's Finches*. Princeton University Press, Princeton.

Grant, B.R. and Grant, P.R. (1989) *Evolutionary Dynamics of a Natural Population: The Large Cactus Finch of the Galápagos*. University of Chicago Press, Chicago.

Grant, P.R. and Grant, B.R. (1992) *Ecology* **73**, 766–784.

Grant, P.R., Grant, B.R., Keller, L.F. and Peteren, K. (2000) *Ecology* **81**, 2442–2457.

Greenwood, J.J.D. and Baillie, S.R. (1991) *Ibis* **133(suppl. 1)**, 121–133.

Houghton, J.T., Ding, Y., Griggs, D.J., Noguer, M., Van der Linden, P.J. and Xiasou, D. (2001) *Climate Change 2001*. Cambridge University Press, Cambridge.

Hurrell, J.W. (1995) *Science* **269**, 676–679.

Hurrell, J.W., Kushnir, Y., Ottersen, G. and Visbeck, M. (2003) In: *The North Atlantic Oscillation* (Ed. by J.W. Hurrell, Y. Kushnir, G. Ottersen and M. Visbeck), pp. 1–35. American Geophysical Union, Washington, DC.

Jones, P.D., New, M., Parker, D.E., Martin, S. and Rigor, I.E. (1999) *Rev. Geophys.* **37**, 173–197.

Kostrzewa, R. and Kostrzewa, A. (1991) *Auk* **108**, 342–347.

Krüger, O. and Lindström, J. (2001) *J. Anim. Ecol.* **70**, 173–181.

Lack, D. (1954) *The Natural Regulation of Animal Numbers*. Clarendon Press, Oxford.

Lack, D. (1966) *Population Studies of Birds*. Oxford University Press, Oxford.

Lande, R., Engen, S. and Sæther, B.-E. (1995) *Am. Nat.* **145**, 728–745.

Lande, R., Engen, S. and Sæther, B.-E. (1999) *Am. Nat.* **154**, 271–281.

Lande, R., Engen, S., Sæther, B.-E., Filli, F., Matthysen, E. and Weimerskirch, H. (2002) *Am. Nat.* **159**, 321–332.

Lande, R., Engen, S. and Sæther, B.-E. (2003) *Stochastic Population Dynamics in Ecology and Conservation*. Oxford University Press, Oxford.

Lebreton, J. (1990) In: *Population Biology of Passerine Birds* (Ed. by J. Blondel, A. Gosler, J.-D. Lebreton and R.H. McCleery), pp. 89–102. Springer-Verlag, Berlin.

Lebreton, J. and Clobert, J.C. (1991) In: *Bird Population Studies* (Ed. by C.M. Perrins, J.-D. Lebreton and G.J.M. Hirons), pp. 105–125. Oxford University Press, Oxford.

Lewontin, R.C. and Cohen, D. (1969) *Proc. Natl. Acad. Sci. USA* **62**, 1056–1060.

Liley, D. (1999). Predicting the consequences of human disturbance, predation and climate change on ringed plover populations. Unpublished PhD Thesis, University of East Anglia.

Loison, A., Sæther, B.-E., Jerstad, K. and Røstad, O.W. (2002) *J. Appl. Stat.* **29**, 289–304.

Lusk, J.J., Guthery, F.S., George, R.R., Peterson, M.J. and DeMaso, S.J. (2002) *J. Wildl. Manage.* **66**, 1040–1051.

Mann, M.E., Bradley, R.S. and Hughes, M.K. (1998) *Nature* **392**, 779–787.

Marchant, J.M. (1992) *Ibis* **134(suppl. 1)**, 113–119.

Martin, T.E. (2001) *Ecology* **82**, 175–188.

Møller, A.P. (1989) *J. Anim. Ecol.* **58**, 1051–1063.

Møller, A.P. (2002) *J. Anim. Ecol.* **71**, 201–210.

Møller, A.P. and Erritzøe, J. (2003) *Oecologia* **137**, 621–626.

Moss, R. (1986) *Ibis* **128**, 65–72.

Moss, R. and Oswald, J. (1985) *Ornis Scand.* **16**, 229–238.

Moss, R., Oswald, J. and Baines, D. (2001) *J. Anim. Ecol.* **70**, 47–61.

Mysterud, A., Yoccoz, N.G., Stenseth, N.C. and Langvatn, R. (2000) *J. Anim. Ecol.* **69**, 959–974.

O'Connor, R.J. (1980) *Ardea* **68**, 165–183.

O'Connor, R.J. (1981) In: *Animal Migration* (Ed. by D.J. Aidley), pp. 167–195. Cambridge University Press, Cambridge.

Orell, M. (1989) *Ibis* **113**, 112–127.

Paradis, E., Baillie, S.R., Sutherland, W.J. and Gregory, R.D. (2000) *Ecology* **81**, 2112–2125.

Parker, D.E., Jones, P.D., Folland, C.K. and Bevan, A. (1994) *J. Geophys. Res.-Atmos.* **99**, 14373–14399.

Persson, C. (1987) *J. Zool. Lond. (B)* **1**, 671–691.

Pimm, S.L. (1991) *The Balance of Nature?* University Chicago Press, Chicago.

Podoler, H. and Rogers, D. (1975) *J. Anim. Ecol.* **44**, 85–114.

Potts, G.R. (1986) *The Partridge*. Collins, London.

Quinn, T.J. and Deriso, R.B. (1999) *Quantitative Fish Dynamics*. Oxford University Press, New York.

Ranta, E., Kaitala, V. and Lindström, J. (1998) In: *Modeling Spatiotemporal Dynamics in Ecology* (Ed. by J. Bascompte and R.V. Solé), pp. 47–62. Springer-Verlag, Berlin.

Reynolds, C.M. (1979) *Bird Study* **26**, 7–12.

Ringsby, T.H., Sæther, B.-E., Tufto, J., Jensen, H. and Solberg, E.J. (2002) *Ecology* **83**, 561–569.

Rivera-Milán, F.F. and Schaffner, F.C. (2002) *Condor* **104**, 587–597.

Rolley, R.E., Kubisiak, J.F., Paisley, R.N. and Wright, R.G. (1998) *J. Wildl. Manage.* **62**, 917–924.

Rotella, J.J., Ratti, J.T., Reese, K.P., Taper, M.L. and Dennis, B. (1996) *J. Wildl. Manage.* **60**, 817–825.

Royama, T. (1992) *Analytical Population Dynamics.* Chapman & Hall, London.

Sæther, B.-E. and Engen, S. (2002) *Phil. Trans. R. Soc. Lond. B* **357**, 1185–1196.

Sæther, B.-E., Ringsby, T.H. and Røskaft, E. (1996) *Oikos* **71**, 273–278.

Sæther, B.-E., Tufto, J., Engen, S., Jerstad, K., Røstad, O.W. and Skåtan, J.E. (2000a) *Science* **287**, 854–856.

Sæther, B.-E., Engen, S., Lande, R., Arcese, P. and Smith, J.N.M. (2000b) *Proc. R. Soc. Lond. B* **267**, 621–626.

Sæther, B.-E., Engen, S. and Matthysen, E. (2002a) *Science* **295**, 2070–2073.

Sæther, B.-E., Lande, R., Engen, S., Both, C. and Visser, M. (2002b) *Oikos* **99**, 331–337.

Sæther, B.-E., Engen, S., Møller, A.P., Matthysen, E., Adriansen, F., Fiedler, W., Leivits, A., Lambrechts, M.M., Visser, M.E., Anker-Nilssen, T., Both, C., Dhondt, A.A., McCleery, R.H., McMeeking, J., Potti, J., Røstad, O.W. and Thomson, D. (2003) *Proc. R. Soc. Lond. B*, 1.

Sæther, B.-E., Engen, S., Møller, A.P., Weimerskirch, H., Visser, M.E., Fiedler, W., Matthysen, E., Lambrechts, M.M., Badyaev, A., Becker, P.H., Brommer, J.E., Bukacinski, D., Bukacinska, M., Christensen, H., Dickinson, J., du Feu, C., Gehlbach, F.R., Heg, D., Hötker, H., Merilä, J., Nielsen, J.T., Rendell, W., Robertson, R.J., Thomson, D.L., Török, J. and Van Hecke, P. *Am. Nat.* (in press a).

Sæther, B.-E., Engen, S., Møller, A.P., Visser, M.E., Matthysen, E., Fiedler, W., Lambrechts, M.M., Becker, P.H., Brommer, J.E., Dickinson, J., du Feu, C., Gehlbach, F.R., Merilä, J., Rendell, W., Robertson, R.J., Thomson, D.L. and Török, J. *Ecology* (in press b).

Sætre, G.-P., Post, E. and Král, M. (1999) *Proc. R. Soc. Lond. B* **266**, 1247–1251.

Sanz, J.J. (2002) *Global Change Biol.* **8**, 1–14.

Sanz, J.J. (2003) *Ecography* **26**, 45–50.

Sanz, J.J., Potti, J., Moreno, J., Merino, S. and Frías, O. (2003) *Global Change Biol.* **9**, 1–12.

Schreiber, R.W. and Schreiber, E.A. (1984) *Science* **225**, 713–716.

Sillett, T.S., Holmes, R.T. and Sherry, T.W. (2000) *Science* **288**, 2040–2042.

Slagsvold, T. (1975) *Norw. J. Zool.* **23**, 67–88.

Slagsvold, T. and Grasaas, T. (1979) *Ornis Scand.* **10**, 37–41.

Steenhof, K., Kochert, M.N., Carpenter, L.B. and Lehman, R.N. (1999) *Condor* **101**, 28–41.

Stenning, M.J., Harvey, P.H. and Campbell, B. (1988) *J. Anim. Ecol.* **57**, 307–317.

Stenseth, N.C., Mysterud, A., Ottersen, G., Hurrell, J.W., Chan, K.S. and Lima, M. (2002) *Science* **297**, 1292–1296.

Stokke, B., Møller, A.P., Sæther, B.-E., Rheinwald, G. and Gutscher, H. *Auk* (in press).

Stott, P.A. and Kettleborough, J.A. (2002) *Nature* **416**, 723–726.

Sutherland, W.J. (1996) *Proc. R. Soc. Lond. B* **263**, 1325–1327.

Szép, T. (1995) *Ibis* **137**, 162–168.

Tett, S.F.B., Stott, P.A., Allen, M.R., Ingram, W.J. and Mitchell, J.F.B. (1999) *Nature* **399**, 569–572.

Thompson, P.M. and Ollason, J.C. (2001) *Nature* **413**, 417–420.

Turchin, P. (1995) In: *Population Dynamics* (Ed. by N. Cappucino and P.W. Price), pp. 19–40. Academic Press, New York.

Verner, J. and Purcell, K.L. (1999) *Condor* **101**, 219–229.

Visser, M.E., Adriaensen, F., van Balen, J., Blondel, J., Dhondt, A.A., van Dongen, S., du Feu, C., Ivankina, E.V., Kerimov, A.B., de Laet, J., Matthysen, E., McCleery, R., Orell, M., and Thomson, D.L. (2003) *Proc. R. Soc. Lond. B* **270**, 367–372.

von Haartman, L. (1973) *Lintumies* **8**, 7–9.

Watson, A., Moss, R. and Rae, S. (1998) *Ecology* **79**, 1174–1192.

Watson, A., Moss, R. and Rothery, P. (2000) *Ecology* **81**, 2126–2136.

Wilson, P.R., Ainley, D.G., Nur, N., Jacobs, S.S., Barton, K.J., Ballard, G. and Comiso, J.C. (2001) *Mar. Ecol. Prog. Ser.* **213**, 301–309.

Yalden, D.W. and Pearce-Higgins, J.W. (1997) *Bird Study* **44**, 227–234.

Importance of Climate Change for the Ranges, Communities and Conservation of Birds

KATRIN BÖHNING-GAESE* AND NICOLE LEMOINE

I. SUMMARY

In this review, we focus on the effects of global climate change on the size and position of geographic ranges and the richness and composition of bird communities. Plenty of evidence exists demonstrating that range boundaries of birds are correlated with climatic factors. In general, the northern range limit of species seems to be influenced rather by abiotic factors such as cold temperatures. The southern range limit of species appears to be determined by climatic factors such as heat or lack of water in arid regions and by biotic factors in more humid regions. For communities, species richness is best predicted by measures of ambient temperature at high latitudes and by water-related variables in low-latitude, high-temperature regions. Models predicting

---

E-mail address: boehning@oekologie.biologie.uni-mainz.de (K. Böhning-Gaese)

ADVANCES IN ECOLOGICAL RESEARCH, VOL. 35
0065-2504/04 $35.00 DOI 10.1016/S0065-2504(04)35010-5

range changes under climate change show idiosyncratic responses of different species with range contractions being more frequent than range expansions. Range shifts have been observed in temperate regions with northward shifts of northern range boundaries and no consistent trend of southern range boundaries. Further, upslope movements have been observed on a tropical mountain. For communities, increases in species richness are predicted for northern latitude and high-elevation sites and declines of species richness in arid regions. With increasing winter temperature, declines in the proportion of migratory species in bird communities have been predicted and observed. Conservation consequences of global climate change are especially high threats to species in arid environments, expected movements of species out of protected areas and increasing land use conflicts. In general, surprisingly few studies document effects of climate change on birds' ranges and communities. Given that range contractions and declines of species richness often initiate conservation efforts, further studies are urgently needed.

II. INTRODUCTION

Recent climate change has been shown to have significant influence on numerous plant and animal species in North America and Europe (Grabherr *et al.*, 1994; Parmesan, 1996; Hill *et al.*, 1999, 2002; Parmesan *et al.*, 1999, 2000; Kirschbaum, 2000; Cameron and Scheel, 2001; Walther *et al.*, 2001; Warren *et al.*, 2001; Bale *et al.*, 2002; Shine *et al.*, 2002; Konvicka *et al.*, 2003; Parmesan and Yohe, 2003; Root *et al.*, 2003). Especially, compelling evidence has been found for the impact of global climate change on the phenology, breeding biology and population dynamics of birds. One important open question is if global climatic change has consequences for the geographic ranges of birds and bird communities. Range contractions, local extinctions and declines in species richness are three of the most frequently used indicators of changing environments that often initiate conservation action. Thus, it is of special interest to search for potential effects of global climate change on ranges and communities. If such effects are found, this might lead to modifications in conservation policy and practice to anticipate and to deal with the most pressing effects of global climate change on birds.

Changes in climate should lead to changes in the geographic ranges of birds. In a changing climate species can respond by: (1) evolutionary adaptation, i.e., true evolutionary change, (2) phenotypic adaptation, i.e., phenotypic plasticity, (3) movement, and (4) extinction (Peterson *et al.*, 2001). The result can be changes in geographic ranges, with possible differences between changes in breeding and non-breeding ranges. For example, if a population of short-distance migratory birds gradually evolves lower levels of migratory restlessness, this would lead to increasing numbers of individuals staying on the wintering grounds and, simultaneously, a possible extension of the wintering range (Berthold, 2000).

Through changes in species' geographic ranges, changes in climate can also lead to changes in species richness of bird communities. Changes in species richness have to be studied explicitly because effects of climate change on the ranges of individual bird species do not necessarily lead to changes in species richness (Lennon *et al.*, 2000). For example, if temperature increases at a site, the area might lose species whose southern range boundaries move north, but one might gain southern species. It could be possible to predict the range changes of the individual species using, for example, the Genetic Algorithm for Rule-set Prediction (GARP, Stockwell and Noble, 1992; Stockwell and Peters, 1999). However, the interactions of a species with its habitat, resources, competitors, predators and parasites might change as well and make predictions unreliable (Schwartz, 1992; Davis *et al.*, 1998). Thus, species richness might decline, remain stable or even increase depending on the net effect of local extinctions and colonisations.

Furthermore, climate can be expected to lead to changes in the composition of bird communities. For example, increases in winter temperature but not in temperature during the breeding season should lead to colonisation of resident species and local extinctions of migratory species with a corresponding decline in the proportion of migratory species within the community (Lemoine and Böhning-Gaese, 2003). In this chapter we, therefore, focus on the effects of global climate change on the size and position of geographic ranges and the richness and composition of bird communities. We then use these results to evaluate consequences for the conservation of birds.

III. CHANGES IN CLIMATE

Changes in climate that influence birds' ranges and communities are, first, increases in mean annual temperature. Warming occurred most rapidly during the periods 1925–1944 and 1978–1997 (Jones *et al.*, 1999). In all probability, warming did not stop in 1997, but newer data were not included in the cited reference. Ranges of species, however, are potentially influenced more by minimum than by average temperatures. Recent analyses of the changes in minimum and maximum temperatures demonstrated that annual minimum temperatures increase more strongly than annual maximum temperatures (Easterling *et al.*, 1997; Currie, 2001). This leads to a decrease in intra-annual seasonality. Furthermore, temperature changes are spatially heterogeneous with higher temperature increases expected at higher latitudes. In addition, temperature changes seem to differ between continents. While Eurasia has shown an overall warming trend since the early 1970s, North America exhibits warming at a lower rate and even a slight cooling trend during the last 50 years in the eastern United States (Bogaert *et al.*, 2002).

Land surface precipitation is another important factor influencing birds' ranges and communities. Precipitation increased about 0.5–1% per decade since the late 1960s in mid- to high latitudes, but showed a decadal decrease of 0.3% in the tropics and subtropics (IPCC, 2001). Further, extreme weather events like extremely cold winters or long droughts are known to influence species' ranges and have been demonstrated to increase in frequency and intensity (Easterling *et al.*, 2000). Finally, pronounced effects on birds' ranges and communities are expected through indirect effects of global climate change, in particular, changes in fire regimes, vegetation, and land use. As one of the most severe threats, one can consider rising sea levels, which might lead to severe habitat loss in coastal areas, with the heaviest impact on intertidal habitat, salt marshes and sandy beaches (Galbraith *et al.*, 2002; Kont *et al.*, 2003). Theoretically, it appears possible that rising sea levels also generate new habitats. However, in most coastal regions, e.g., the Wadden Sea, the coastline is "fixed" by artificial dikes.

IV. HOW TO ANALYSE THE EFFECT OF CLIMATE CHANGE

While it is not difficult to delineate theoretically the consequences climate change has on birds' ranges and communities, it is a great challenge to actually demonstrate a causal relationship between changes in climatic factors and, e.g., northward extension of a species' range. A number of different approaches have been taken:

(1) A traditional approach has been to correlate the range of a species, particularly the range boundary with a climatic variable over a latitudinal or an elevational gradient (Root, 1988). Similarly, species richness or community composition has been correlated with climatic factors (MacArthur, 1972; Currie, 2001). This approach demonstrates only that climatic factors are directly or indirectly, involved in influencing ranges and communities. The approach does not show that climate change *causes* changes in ranges and communities.

(2) In a second approach, the correlation between climatic factors on the one hand and species' ranges and communities on the other hand has been used to predict changes in the face of expected climatic change. In most studies, investigators established the relationship between, e.g., the presence of a species or between species richness in a grid cell

and the climatic factors of this grid cell, then used global circulation models (GCM) to calculate expected changes in the climatic factors in the respective grid cell and, finally, predicted the change in the presence of the species or in species richness (Currie, 2001; Peterson *et al.*, 2001, 2002; Peterson, 2003a). This is the only method to gain an understanding about the magnitude that potential effects of climate change have on ranges and communities. The method, however, has a number of limitations: (a) It depends on the accuracy and resolution of the GCMs (Currie, 2001; Peterson *et al.*, 2001). (b) In a number of studies, predicted changes in climate lead to temperature or precipitation levels that were outside the range of variables in the contemporary data set. Predicted values for presence of a species or species richness in the respective areas may therefore be suspect (Currie, 2001). (c) The studies about range changes, conducted so far, were limited to a politically defined region, e.g., Mexico. Species that are predicted to move their ranges outside this region will be noted as going locally extinct. It is, however, not possible to predict which species might expand the ranges into Mexico and colonise the region. This leads to a bias towards losses when predicting changes in species richness (Peterson *et al.*, 2002). The range models assume that present range boundaries are controlled only by climatic factors. Changes in biotic interactions might make predictions unreliable (Davis *et al.*, 1998; Hulme, 2003; Pearson and Dawson, 2003).

(3) In a third approach, observed changes in the ranges of species or in communities have been attributed to global climate change. For example, in the presence of a warming climate, British birds showed a mean northward shift of the northern range boundary with no systematic shift of the southern range boundary (Thomas and Lennon, 1999). In these studies, the authors have accumulated evidence that the observed changes are caused by climate change. For example, the changes conformed to a model that predicts how strong the change should be (given, e.g., the spatial relationship between climatic factors and species richness and given the observed climatic change; Lemoine and Böhning-Gaese, 2003). Other authors used as evidence that the site under study has experienced climatic change, that the range boundary of the species under study correlates with a climatic factor, or that the species under study shows a fitting physiological temperature tolerance towards high or low temperatures (Thomas and Lennon, 1999). Furthermore, some authors tested explicitly alternative factors. For example, Thomas and Lennon (1999) showed that the northward extension of a species' range is not correlated with a general increase in the abundance and range size of the species.

V. INFLUENCE OF CLIMATE CHANGE ON THE RANGES OF BIRDS

In this and the following section, we present evidence that birds' ranges and communities are influenced by climate change. The structure of the two sections follows, in general, the three approaches described above. First, we assess evidence that the current ranges of species and the current richness and composition of bird communities are correlated with climatic factors. We use these correlations to make predictions that changes should occur under climate change. Second, we present the results of modelling approaches that analyse the potential impact of climate change on ranges and communities. Third, we review the literature about actually observed influences of global climate change on ranges and communities. Finally, we evaluate whether the changes we predicted were confirmed by the modelling and observational studies.

A. Correlations of Climatic Factors and Range Boundaries

For birds, plenty of evidence exists demonstrating that range boundaries are correlated with climatic factors (for reviews, see Brown and Lomolino (1998), Gaston (2003) and Newton (2003)). In a classic study, for example, Root (1988) demonstrated that 62 species of North American passerines and non-passerines have a northern range boundary that was associated with a particular January isotherm, with different species associated with different isotherms. Root (1988) suggested that the northern range limit is influenced by physiological constraints. This is supported by the fact that the metabolic rate of the species at the January temperature prevailing at the northern boundary was calculated at about 2.5 times the basal metabolic rate (but see Repasky (1991), Root and Schneider (1993)). Other factors that are correlated with climatic factors and that might indirectly determine the northern range boundary are day length (influencing the time period available for foraging), and the presence of abundant, energy-rich food resources. In addition, northern range limits of breeding distributions might be influenced by high wetness as in Britain, for example. There, the capercaillie *Tetrao urogallus* may be limited in distribution by high rainfall, which leads to poor chick survival (Moss, 1986).

In contrast, evidence exists that the southern range boundary of birds can be associated with heat or lack of water. The Black-billed Magpie *Pica pica* in North America, for example, has a heat limit of 40°C, while the related yellow-billed magpie *Pica nuttali* is more heat tolerant. Correspondingly, the Black-billed Magpie is confined to cooler areas, while the yellow-billed magpie can live in hotter areas (Hayworth and Weathers, 1984). The fact that lack of drinking water can limit bird distribution is shown by the failure of most species to penetrate far into deserts. When water is made available by humans, e.g., by providing drinking sites for cattle and sheep as in Australia, this can lead to range expansions of birds (Saunders and Curry, 1990). Again, rainfall can act on birds also indirectly via the food supply of insects, flowers, and seeds (Newton, 2003).

In general, the northern range limit of species seems to be influenced largely by abiotic factors such as cold temperatures or high wetness (MacArthur, 1972; Newton, 2003). The southern range limit of species might be influenced, on the one hand, by climatic factors such as heat or lack of water in regions in which they are limiting factors, e.g., in arid regions (Newton, 2003). On the other hand, in more humid regions, biotic factors such as interspecific competition, predation, and parasitism might play a more pronounced role (Dobzhansky, 1950; MacArthur, 1972; Brown and Lomolino, 1998; Hofer et al., 1999; Gross and Price, 2000).

Given the observed changes in climate and our current knowledge about factors influencing range boundaries we can make predictions about the consequences of global climate change. With increasing annual temperatures, especially increasing minimum temperatures and increasing temperatures especially at high latitudes, we would expect northward range expansions of birds at high altitudes and in temperate, boreal, and Arctic regions. In contrast, with increasing temperatures and decreasing precipitation in subtropical and tropical regions, we would expect range contractions in the respective areas. Certainly, these predictions have to be treated with great care because indirect interactions between abiotic and biotic factors can lead to unexpected patterns, especially when trying to make forecasts for individual species. Particularly difficult to anticipate are changes in southern range boundaries in cases in which abiotic factors are not limiting and in which range boundaries are determined by a complex interplay of biotic factors or by certain vegetation types which might themselves be influenced by climate change (Thomas and Lennon, 1999).

B. Modelled Consequences of Climate Change for Range Boundaries

A pivotal study that analysed the potential effects of global climate change on birds' range, using modelling approaches, was conducted by Peterson et al. (2001) (Table 1; Figure 1). They described the current distribution of eight Mexican Cracidae species with an ecological niche model relating the presence of each species at a concurrent site to the environmental variables. They then used two GCMs, one conservative and one less conservative model, to calculate the climatic change during the next 50 years. For each climate scenario, they evaluated the expected change in species' distributions under three assumptions regarding different dispersal abilities of the species (no dispersal, dispersal to contiguous areas and universal dispersal) and compared them with the current distribution of each species. Based on the two different scenarios of climate

Table 1 Modelled changes in the ranges and communities of birds attributed to future climate change

References	Variable modelled	Species modelled	Change	Method	Region
Zöckler and Lysenko (1994)	Breeding area	23 Arctic bird species	Loss of breeding habitat, decrease of almost 50% of total geese population	MAPSS model, HadCM2GSa1 model, UKMO model, $CO_2 \times 2$, 4–7° temperature increase	Tundra of the northern hemisphere
Peterson et al. (2001)	Geographic distribution	8 cracid species	Radical reductions to moderate increases	GARP model, 2 GCMs, 3 dispersal abilities, $CO_2 \times 2$	Mexico
Benning et al. (2002)	Elevational range of disease vector	10 forest birds	Range contractions of the 10 species of endangered forests birds	30 m US Geological Survey digital elevation model, 2°C temperature increase	Hawaiian Islands
Peterson (2003a)	Geographic range	26 Rocky Mountain birds, 19 Great Plain birds	Range shifts of montane birds smaller than for flatland species	GARP model, 2 GCMs, 3 dispersal abilities, $CO_2 \times 2$	Central and western North America
Erasmus et al. (2002)	Geographic range	34 bird, 19 mammal, 50 reptile, 19 butterfly, 57 "other invertebrate" species	Some range expansions, most range contractions, western movement towards eastern highlands	GCM, $CO_2 \times 2$, 2°C temperature increase	South Africa
Currie (2001)	Species richness	Birds, mammals, reptiles, amphibians, trees	Decrease in bird richness, increase in cooler high-elevation areas, positive effects for reptiles and amphibians	Polynomal regression model of species richness, mean January and July temperature and precipitation, 5 GCM, $CO_2 \times 2$	Conterminous USA
Peterson et al. (2002)	Geographic distribution, species richness	1179 birds, 416 mammals, 175 butterflies	Colonisations and increase for major sierras, extinctions and loss for Chihuahuan desert and costal plain	GARP model, 2 GCMs, 3 dispersal abilities, $CO_2 \times 2$	Mexico

218

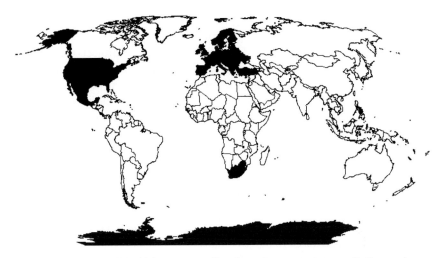

Figure 1   Countries in which recent studies about the consequences of climate change for range boundaries or community composition of birds were conducted.

change and the three assumptions concerning species' dispersal abilities, the effects on species distributions show a great variety of patterns from radical reduction to moderate increases. These idiosyncratic changes of range boundaries for the different species indicate that the effects of climatic change can differ very much, depending on species, GCM, and dispersal scenario.

To find more general patterns in predicted range changes, Peterson (2003a) (Table 1; Figure 1) compared predicted changes of montane and lowland species. He expected to find that high-altitude species are especially threatened by climate change because increasing temperatures can reduce high-altitude habitat areas with associated range contractions and potential loss of high-altitude species (McDonald and Brown, 1992). He compared expected range changes of 26 species restricted to the Rocky Mountains and 19 species restricted to the Great Plains using the same general approach as in Peterson *et al.* (2001). The results, however, did not fit the expectations. The predicted range reductions and range shifts of montane species were smaller than for the lowland species. These results were attributed to the fact that much more dramatic horizontal shifts would be expected in lowland systems simply because of the geometry of the landscape, given the same intensity of temperature shifts. However, predictions might be different for highest montane species that were not included in the data set (Peterson, 2003a).

In contrast to these marginal effects on high-altitude species, more negative consequences were predicted for high-latitude, Arctic species. Zöckler and Lysenko (1994) (Table 1; Figure 1) correlated the current distribution of 23 Arctic water bird species with the current vegetation, and then modelled the likely changes in the vegetation to investigate the impacts that this would have on species' distributions. For mapping vegetation changes, the mapped plant soil system (MAPSS) equilibrium biogeography model was used, which links vegetation to climate. Two GCMs were then used to project the climate conditions with CO_2 doubling, which predicts a temperature increase from 4 to 7°C for Arctic latitudes. The majority of Arctic water birds spend their breeding season in areas of tundra vegetation. Changes in vegetation patterns will be most extreme for tundra areas. With climate warming, it is expected that taiga and boreal forests expand their distribution towards the north. Overall losses of current tundra distribution are estimated between 40 and 57%, with most being replaced by forest. New tundra areas amount to only 5%. Assuming that habitat area is a limiting factor, and that birds are evenly distributed within their ranges, the predicted change in vegetation could mean, e.g., a loss of almost half of the total goose population.

While the previous studies analysed the impact of climate change on the upper, altitudinal and northern range boundary of birds, only one study described potential effects on the lower, altitudinal range boundary. In this study, Benning et al. (2002) (Table 1; Figure 1) used a modelling approach to predict the consequences global warming has on the distribution of a disease vector and connected changes in the distribution of birds. On the Hawaii islands, the distribution of many endemic Hawaiian bird species is limited by the distribution of the introduced mosquito, Culex quinquefasciatus, and the avian malaria that it transmits. The native species are restricted to high elevation forests because they are sensitive to avian malaria, and because the disease vector occurs only at lower elevations. With climate change, the elevational range of Plasmodium is expected to increase. Assuming a 2°C increase in temperature, which should occur by some time late in the current century, the forest area with intermediate and low risk of malaria infection is expected to decrease dramatically. This should lead to threatening range contractions of the 10 species of endangered forests birds that are now restricted to these high-elevation forests.

C. Observed Consequences of Climate Change for Range Boundaries

Actual range changes of birds are documented in many studies (for a summary, see Newton, 2003, p. 425. However, in only the following cases have these range changes been explained by climate change. In two first studies, Kalela (1949, 1952) (Table 2; Figure 1) found changes in the distribution for a few mammal and several bird species in Germany and Finland since the 1850s, which he could connect with corresponding changes in winter, spring and/or summer temperature or the desiccation of a species' breeding area. But, as Kalela (1949, 1952) himself pointed out, range expansions, especially in southern Finland, can also be caused by cultural factors. Forest clearings, for example, have created huge areas of new habitat for open country species.

Table 2 Observed changes in the ranges and communities of birds attributed to recent climate change

References	Variable observed	Species observed	Change	Cause	Time span	Region
Kalela (1949)	Northern range expansions	37 species	Northwards expansion of 17 species	Present climatic changes, increased atmospheric circulation, activities of man	1850–1950	Southern Finland and Germany
Kalela (1952)	Geographic distribution	3 mammals, 38 bird species	Northwards expansion	Increasing temperature or drying-up of lakes	1850–1950	Finland
Johnson (1994)	Nesting distribution	24 bird species	Shift of nesting range mainly in northern direction	Increased summer moisture, perhaps higher mean temperature	1957–1992	Western United States
Burton (1995)	Range expansions	424 bird species	195 advancing to north, 64 advancing to west, 56 retreating to south, 7 retreating to east	Climate warming or cooling	1900–1995	Europe
Thomas and Lennon (1999)	Geographic range	59 southern bird species, 42 northern bird species	18.9 km northwards shift of northern boundary	Increasing temperatures	1968–72 to 1988–91	UK
Peterson (2003b)	Geographic distribution	5 bird species	3 species northward shifts, 1 species southward shift	Global warming	Last 50 years	Great Plains of North America
Pounds et al. (1999)	Upper range boundary along elevational gradient	Abundance of birds	Increases in abundance and colonisation events of premontane species	Increase in sea surface temperatures in the equatorial Pacific, altitudinal rise of orographic clouds	1979–1998	Monteverde, Costa Rica
Lemoine and Böhning-Gaese (2003)	Percentage long-distance migratory species	151 landbird species	1.76% decline of proportion of long-distance migrants	Warmer winter temperature	1980–1981 to 1990–1992	Lake Constance, central Europe

221

Similar shifts were observed by Burton (1995) in a European-wide study. He found spectacular changes in the breeding ranges of birds since 1900 with differences in the direction in which distributions shifted. He recorded a northwards advance of 195 and a westwards advance of 64 temperate bird species due to climatic warming mostly between 1900 and 1950. At the same time, 32 species showed a retreat towards the southern parts and 7 species a retreat towards the eastern parts of their ranges. Since 1950, 24 temperate bird species retreated to the southern parts of their ranges because of climate cooling. For northern breeding bird species the pattern was quite different. For 52 species, Burton (1995) found a retreat towards the northern parts of their breeding range and for nine species a southward advance of ranges in response to climate warming. A southward advance of 55 northern breeding birds could be attributed to climatic cooling.

In the most recent European study, Thomas and Lennon (1999) (Table 2; Figure 1) analysed the northern and southern edge of the range of birds in Great Britain. They were the first to use a rigorous statistical, community-wide approach. They used atlas data of 1968–1972 and 1988–1991 and controlled for general range expansions and contractions of the species. They found that the northern range boundaries of 59 southern British bird species moved on average 18.9 km towards the north. Southern range boundaries of 42 northern bird species did not change systematically over the same time period. The authors claim that the most parsimonious explanation for these results was climatic change because recent changes in timing and breeding success of birds were correlated with spring temperatures (Crick et al., 1997; Visser et al., 1998), the spatial distribution of British bird diversity was connected with temperature (Turner et al., 1988), summer temperature was a significant predictor of breeding distribution of 45% of terrestrial and freshwater bird species (Lennon et al., 2000), and the observed shifts in range boundaries were concordant with a period of climate warming.

Range shifts were observed not only for European but also for North American bird species. In the western United States Johnson (1994) (Table 2; Figure 1) observed changes for 24 bird species that expanded their breeding distribution over the last 30 years. Fourteen of them shifted northwards, four moved in the opposite direction, five species shifted in western direction and one species showed radial expansion. He attributed these range expansions to increased summer moisture, perhaps coupled with a higher mean temperature in the region.

In a recent study, Peterson (2003b) (Table 2; Figure 1) found changes in the geographic distributions of five bird species endemic to the Great Plains of North America. Three species showed significant or near-significant northward shifts, and one a significant shift southward. Of the five species examined overall, colonisation events were focused in the northern half of the distributions of the species (5 of 5 species), whereas extinctions tended to be in the southern half of distribution of the species (3 of 5 species). Peterson (2003b) concludes that the changes in distribution have been subtle and might be associated with global climate change.

In addition, it is expected that the, in some regions dramatic, changes in the abundance of southern temperate and Arctic seabird species (Moller *et al.*, 2004, this volume) also lead to changes in their ranges. For example, at least eight species formed new breeding locations well to the south of their historical range and/or have seen marked population increases at their more southerly colonies since the 19th century on a number of islands off the coast of Western Australia (Dunlop, 2001). Similarily, Ainley *et al.* (1994) observed that colonies of the Adelie penguin *Pygoscelis adeliae* declined and disappeared in the northern- and increased in the southern-most part of the species range during the last 50 years. However, for these species it is difficult to connect range changes with climate change because the populations are heavily influenced by El Niño Southern Oscillation activity which might itself change in frequency and intensity by climate change (Dunn, 2004, this volume).

There were some anecdotal evidence for range expansions towards higher altitudes in middle Europe, but we found only one study that described recent changes in the upper range boundary along an altitudinal gradient. Pounds *et al.* (1999) (Table 2; Figure 1) studied changes in climatic factors and bird communities in highland forests at Monteverde, Costa Rica. They found, connected with an increase in sea surface temperatures in the equatorial Pacific, a rise of the average altitude at the base of the orographic cloud bank, which leads to a decrease in dry-season mist frequency. Studying the abundances of bird species at 1540 m height between 1979 and 1998, they found significant increases in abundance and colonisation events of premontane, cloud-forest intolerant species. In contrast, there was no consistent trend for lower montane, cloud-forest species. As one result of these changes, keel-billed toucans *Ramphastos sulfuratus*, typical for lowlands and foothills, now nested alongside resplendent quetzals *Pharomachrus moccino*, which symbolise Middle American cloud forests. The changes could not be explained by recent deforestation in the lowlands (Pounds *et al.*, 1999).

Comparing our predictions about the effect of global climate change on ranges with the modelled and actually observed changes presents many consistencies. Despite the fact that predicted and observed changes in range boundaries show great variation among different bird species, general patterns can be found. We expected northward range expansions of birds at high altitudes and in temperate, boreal, and Arctic regions. These range expansions have been predicted for many bird species in modelling approaches (Zöckler and Lysenko, 1994; Peterson *et al.*, 2001; Peterson, 2003a) and in observational studies (Kalela, 1949, 1952; Ainley *et al.*, 1994; Johnson, 1994; Thomas and Lennon, 1999; Pounds *et al.*, 1999; Dunlop, 2001; Peterson, 2003b). In arid subtropical and tropical regions, we expected range contractions with increasing temperatures and decreasing precipitation. This pattern has not been found in the literature reviewed above but evidence exists from the community-wide studies (see Section VI below; Erasmus *et al.*, 2002; Peterson *et al.*, 2002). For the southern latitudinal and the lower elevational range boundary in humid regions we had refrained from predictions because it is usually assumed that these boundaries are more strongly influenced by more complex, biotic interactions. Correspondingly, the southern range boundaries of British birds (Thomas and Lennon, 1999) and the lower, elevational range limits of cloud-forest species on Monteverde (Pounds *et al.*, 1999) did not shift under changing climate. The study on Hawaiian birds (Benning *et al.*, 2002) demonstrated that the lower, elevational range limit of these birds was indeed controlled by a biotic factor, i.e., a disease vector. The disease vector, however, was itself controlled by temperature and was expected to shift its range boundary upwards leading to upward shifts of the lower, elevational range boundary of the birds.

VI. INFLUENCE OF CLIMATE CHANGE ON AVIAN COMMUNITIES

A. Correlations of Climatic Factors and Avian Communities

With respect to avian communities, we will focus on species richness and the composition of bird communities. Also for species richness, plenty of evidence exists for a tight correlation between climatic factors and species richness (reviewed by Begon *et al.* (1996), Gaston (1996) and Brown and Lomolino (1998)). Among the climatic factors, measures of ambient energy (mean annual temperature, potential evapotranspiration (PET), solar radiation) either have a primary role or are an important modulating factor (Turner *et al.*, 1988; Currie, 1991; Lennon *et al.*, 2000; Van Rensburg *et al.*, 2002; Hawkins *et al.*, 2003). For example, Currie (1991) calculating bird species richness on the basis of $2.5 \times 2.5°$ grid cells in northern North America found that PET, but not actual evapotranspiration (AET), was the best predictor of species richness. However, these results might be influenced by the spatial gradient of the analysis.

Other studies that also included also more arid regions demonstrated that water-related variables (AET, annual rainfall) also might play a role. For example, Van Rensburg *et al.* (2002) calculated bird species richness at 0.25, 0.5, and 1° grid cells over South Africa and Lesotho. They found that in this region, the relationship between species richness and PET was unimodal because low rainfall constrained productivity in high-PET areas. In a global analysis of bird species richness at the scale of 220×220 km grid cells, Hawkins *et al.* (2003) demonstrated a latitudinal shift in constraints on diversity. Measures of ambient temperature, e.g., PET, best predicted the diversity gradient at high latitudes, whereas water-related variables, e.g., AET, best predicted richness in low-latitude, high-energy regions.

With respect to the composition of avian communities, most studies looking for large-scale patterns that are correlated with environmental factors have analysed the proportion of migrants in the communities. In an early study, Herrera (1978) demonstrated that the proportion of migratory individuals in forest bird communities increases along a latitudinal gradient in Europe. He showed that the proportion of migrants is correlated with temperature of the coldest month and argued that the proportion of migrants was determined by the difference in carrying capacity between summer and winter, i.e., intra-annual seasonality (Figure 2). Similar patterns were found by Rabenold (1979, 1993), Newton and Dale (1996a,b), Hurlbert and Haskell (2003) and Lemoine and Böhning-Gaese (2003). For example, Lemoine and Böhning-Gaese (2003) correlated the proportion of long-distance migratory bird species in bird communities at the scale of 100×100 km grid cells within Europe (Figure 3) with environmental factors. They found that the proportion of long-distance migratory species increased with increasing spring temperature and decreasing temperature of the coldest month, i.e., with increasing seasonality (Figure 4).

Given the observed changes in climate and our current knowledge about factors influencing bird species richness and community composition we can again make predictions about the consequences of global climate change. With increasing annual temperatures especially at high latitudes, we would expect increases in species richness in temperate, boreal, and Arctic regions. In contrast, with increasing temperatures and decreasing precipitation in arid subtropical and tropical regions, we would expect decreases in species richness in the respective areas. Furthermore, with the expected increases especially in minimum temperatures, and decreasing seasonality, we expect declines in the proportion of migratory individuals and species within communities.

B. Modelled Consequences of Climate Change for Avian Communities

Changes in species richness with climate change have, so far, only been studied using modelling approaches. One type of modelling approach is an extension of the modelling methods used to predict range changes. If range changes for all species in a community can be predicted, it is then possible to add up the number of species per grid cell and to calculate predicted changes in species richness. Using this approach, Peterson et al. (2002) (Table 1; Figure 1) extended the methods used in Peterson et al. (2001) (see Section V) to all bird taxa, all mammal taxa and all butterflies in the families Papilionidae and Pieridae in Mexico. The three taxonomic groups showed similar results. Species showed idiosyncratic changes in ranges with 0.0–2.4% of the species predicted to lose ≥90% of the present distributional area, depending on the dispersal assumption. Colonisations and increases in species richness were predicted for the major

Hard winter conditions

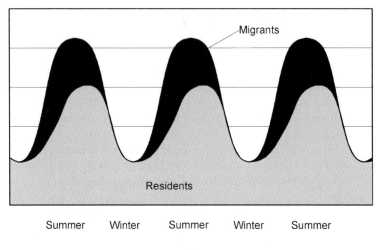

Summer   Winter   Summer   Winter   Summer

Mild winter conditions

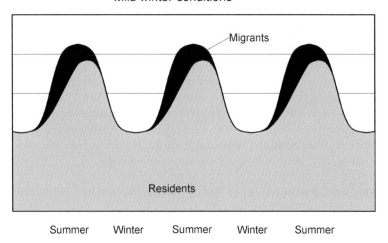

Summer   Winter   Summer   Winter   Summer

Figure 2 Hypothetical fluctuations in resources in the course of the year under seasonal conditions, with severe and mild winters. Grey: resources potentionally used by resident birds; black: resources potentionally available to migrants.

sierras of Mexico. Extinctions and loss of species richness occurred in the broad, open Chihuahuan desert and the north-western coastal plain. Foci of species turnover were northern Mexico in the Chihuahuan desert and the interior valleys in the Baja California peninsula.

Erasmus *et al.* (2002) (Table 1; Figure 1) used a similar approach in South Africa working with 34 bird, 19 mammal, 50 reptile, 19 butterfly, and 57

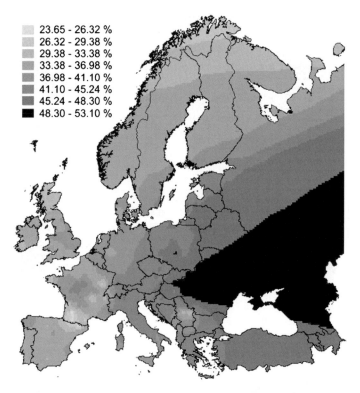

23.65 - 26.32 %
26.32 - 29.38 %
29.38 - 33.38 %
33.38 - 36.98 %
36.98 - 41.10 %
41.10 - 45.24 %
45.24 - 48.30 %
48.30 - 53.10 %

Figure 3   Current proportion of long-distance migratory bird species in European bird communities. The map is based on an interpolation of data from 595 grid cells of 100×100 km^2 (data published in Lemoine and Böhning-Gaese, 2003).

"other invertebrate" species. They applied a multivariate climate envelope approach to describe the current distribution of the species. They then used a GCM simulating a doubling CO_2 concentration that is expected to cause a mean temperature increase of 2°C, and predicted the new distributions of the species. Again, a similar pattern was found for the five different taxonomic groups. The species showed idiosyncratic changes in ranges with some species predicted to experience range expansions but most species to suffer range contractions. Four species (2.2%) were predicted to go locally extinct. Most species showed a western movement and also movement up altitudinal gradients towards the eastern and southeastern highlands. This leads to a predicted concentration of the most species-rich areas on the eastern escarpment, with significant species losses occurring in the western arid regions. In general, the size of species-rich areas was predicted to decline.

A second type of modelling approach was used by Currie (2001) (Table 1; Figure 1). He correlated species richness of birds, mammals, reptiles,

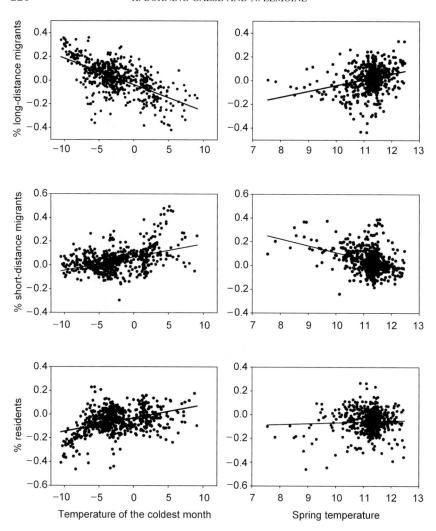

Figure 4 Influence of temperature of the coldest month and temperature in spring (°C) on the proportion of (a) long-distance migrants, (b) short-distance migrants and (c) resident bird species in Europe, after spring precipitation and their squared values were controlled. Lines are regression lines (data published in Lemoine and Böhning-Gaese, 2003).

amphibians, and trees directly with a number of environmental variables in grid cells of $2.5 \times 2.5°$ over the conterminous United States. He then used five different GCMs and predicted, for a doubling CO_2 scenario, changes in environmental variables and, thus, species richness of the different taxonomic groups.

Bird species richness was predicted to be severely affected by climate change because bird richness, more than that of any other group, tended to be lower in hot areas. Predicted higher temperatures over most of the central and southern US should be accompanied by decreases in bird richness, except at higher elevations. Increases in richness are likely to occur in cooler, high-elevation areas, particularly in the western US. The same pattern was found for mammals. Positive effects of predicted climate change was found for reptiles and amphibians, and variable effects for trees.

Various changes in the ranges and community composition of birds have been predicted in the miscellaneous modelling approaches but one has to keep in mind that birds have also the option of evolutionary changes, which does not automatically lead to range changes. Those evolutionary changes were not included in any of the above-mentioned modelling approaches.

C. Observed Consequences of Climate Change for Avian Communities

Only little evidence exists that climate change can lead to changes in the composition of bird communities. In their study of European bird communities, Lemoine and Böhning-Gaese (2003) (Table 2; Figure 1) showed a relationship of the proportion of long-distance migrants, short-distance migrants and resident bird species with spring temperature and temperature of the coldest month (see above, Figure 2). They predicted that the proportion of long-distance migrants should decrease if winter temperature increases and spring temperature does not change (Figure 4). They then used the spatial relationship between bird community structure and climate in Europe to make predictions for changes in the bird communities in the Lake Constance region, Central Europe, between two census periods (1980–1981 and 1990–1992). Winter temperature in this region increased significantly between the two censuses, whereas spring temperature did not change. As predicted from the models the proportion of long-distance migrants decreased and the proportion of short-distance migrants and residents increased between the two censuses. The significant declines of the long-distance migrants in the Lake Constance region were of a magnitude that can be explained by the observed climate change.

Comparing our predictions about the effect of global climate change on communities with the modelled and actually observed changes presents, again, many consistencies. With increasing annual temperatures especially at high latitudes, we predicted increases in species richness in temperate, boreal, and Arctic regions. These predictions were confirmed by three independent, and in their approach very different models for the United States (Currie, 2001), Mexico (Peterson et al., 2002) and South Africa (Erasmus et al., 2002). However, so far, no corresponding studies based on field data have been published. In arid subtropical and tropical regions, we predicted a decrease in species richness with increasing temperatures and decreasing precipitation. Again, these predicted changes were supported in the three modelling studies but up to now, not in the field. Furthermore, we expected declines in the proportion of migratory species within communities with increasing minimum temperatures and decreasing seasonality. Corresponding shifts in community composition were found in the Lake Constance region (Lemoine and Böhning-Gaese, 2003).

VII. CONSERVATION CONSEQUENCES OF CLIMATE CHANGE

No data demonstrating that climate change led to severe population declines or even extinctions of bird species have, to our knowledge, as yet been published. Thus, we have to rely on the predictions derived from the modelling approaches and the observed changes reported so far. In general, we predict and observe range expansions of birds that lead to increases in species richness in many temperate, boreal, and Arctic regions and at high altitudes. The question arises if high Arctic and Alpine species are threatened by these processes because their life-zones might contract. However, models developed by Peterson (2003a) suggest that montane species suffer less range contractions and shifts than flatland species. A problem might pose the highest arctic and alpine species (Zöckler and Lysenko, 1994). In contrast, we predict range contractions and loss of species richness with increasing temperatures and decreasing precipitation in arid subtropical and tropical regions. This demonstrates that the more threatening changes for species and communities are expected in arid subtropical and tropical regions. Thus, conservation efforts have to focus more on the high-temperature, low-precipitation part than on the low-temperature part of the latitudinal gradient.

The most frightening result of all approaches to model range changes are the large movements of ranges expected to occur within the next 50–100 years (Erasmus et al., 2002; Peterson et al., 2002). These range changes differ among species with range contractions being more frequent than range expansions. These changes can lead to a re-shuffling of bird communities with old interactions disappearing and new interactions being created. Climate change might also facilitate colonisation and range expansions of exotic, introduced bird species. For example, recent range expansions of birds, formerly restricted to Africa and tropical Asia such as the Egyptian goose Alopochen aegyptiacus and the rose-ringed parakeet Psittacula krameri in Central Europe have been aided by, among other factors, warmer winter temperatures (Bauer and Bezzel, 2001).

Furthermore, centres of species richness are expected to shift or even to contract (Erasmus *et al.*, 2002). This has severe consequences for conservation efforts. Currently, international conservation efforts focus on areas with many endemic bird species (endemic bird areas, EBAs) or otherwise threatened species (important bird areas, IBAs). With the predicted changes in ranges, it can be expected that many threatened species move out of the areas that receive highest conservation efforts. The new ranges might then be placed in areas with high human population density or areas with conflicting land use (Erasmus *et al.*, 2002).

For example, in their study on Hawaiian birds, Benning *et al.* (2002) demonstrate that not only the disease vector *Culex quinquefasciatus* is expected to shift in elevation. Rather it can be expected that global warming also leads to an increase in the lifting condensation level and the height of orographic clouds. These changes would increase both the lower and upper altitudinal limits of montane cloud forests. Thus, if forests expanded to higher elevations, birds should be able to expand their ranges as well and to escape the areas with high risk of malaria infection. However, in one of the studied forest reserves, Hakalau Wildlife Refuge on the island of Hawaii, the predominant land use above Hakalau Refuge is pasture land, which constrains the amount of forest available at higher elevations and could prevent migration of forests upslope. This demonstrates that under climate change the potential for land-use conflict increases (Erasmus *et al.*, 2002). Thus, in the face of global climate change conservation policy and practice will need to be revised considerably (Berry *et al.*, 2002; Midgley *et al.*, 2002; Hannah *et al.*, 2002).

VIII. CONCLUSIONS AND FUTURE DIRECTIONS

The most interesting conclusion of this review is that so far, only few effects of climate change on ranges and communities have actually been reported using field data. This is surprising because strong effects of climate change on the phenology, reproductive success and population dynamics of birds are documented very well (Lehikoinen *et al.*, 2004; this volume; Dunn, 2004, this volume). This is also surprising because climate has already changed significantly and because expected changes, especially in ranges, are dramatic (Erasmus *et al.*, 2002; Peterson *et al.*, 2002). This result is, furthermore, surprising because birds are one of the few groups of organisms for which good, long-term data exist.

The reasons for this lack of effects are difficult to tell. One possibility is that only very few scientists have looked for effects of climate change on ranges and communities. A second factor might be that changes in ranges and communities are not as strong as we would predict and these ("negative") results have not been published. Finally, it is possible that the regions in which the best data sets exist are not the regions in that climate has changed so far (Figure 1). For example, one of the best data sets with which it should be possible to document changes in the ranges of birds over the last 30 years is the North American Breeding Bird Survey (see Böhning-Gaese *et al.*, 1993; Brown *et al.*, 1995; Villard and Maurer, 1996; Mehlmann, 1997; Husak and Maxwell, 2000; Rodriguez, 2002). However, in contrast to Eurasia, North America exhibits warming at a lower rate and even a slight cooling trend during the last 50 years in the eastern United States (Bogaert *et al.*, 2002).

Contrary to birds, reported changes in ranges and communities are numerous for plants and butterflies in North America and Europe (Grabherr *et al.*, 1994; Parmesan, 1996; Hill *et al.*, 1999; Parmesan *et al.*, 1999, 2000; Kirschbaum, 2000; Cameron and Scheel, 2001; Walther *et al.*, 2001; Warren *et al.*, 2001; Bale *et al.*, 2002; Hill *et al.*, 2002; Shine *et al.*, 2002; Konvicka *et al.*, 2003; Parmesan and Yohe, 2003; Root *et al.*, 2003). Birds are known to react very fast to environmental changes. Thus, no reason exists why birds should respond to climate change less rapidly than plants and butterflies. This suggests that only few scientists have actually looked for effects on birds' ranges and communities. Given the good data sets about birds and the importance of changes in ranges and communities to initiate conservation efforts, this field urgently needs more attention.

Studying range and community changes in birds, it is important to establish as well as possible the relationship between climate and range or community change. One prerequisite is to study not only a single species but also groups of organisms or entire communities (Thomas and Lennon, 1999; McCarty, 2001). Second, it is useful to combine modelling and observational approaches (Lemoine and Böhning-Gaese, 2003). This combination would aid in developing more realistic models, and it would demonstrate in which direction and how strongly individual bird species should shift their ranges. Further, it would be important to collect as much evidence as possible that the ranges of the studied species are limited by climatic factors, e.g., by correlating range boundaries with climatic factors or by studies of physiological temperature tolerance. Finally, it is necessary to explicitly test alternative hypotheses, especially confounding changes in habitat availability and other anthropogenic impacts (Thomas and Lennon, 1999).

In addition, studying bird communities, so far no study has looked beyond bird communities and considered the consequences, that changes in bird abundance and diversity have for the ecosystems within which birds live. Cases where birds have been shown to influence ecosystems are bottom-up and top-down control of food webs. Birds are important disease vectors, flower pollinators, and seed dispersers. For example, seabirds can transport considerable amounts of energy and nutrients from the sea onto islands and shape whole island food webs (e.g., Moreira, 1997; Stapp *et al.*, 1999). Top-down control of food webs can act through grazing or predation in terrestrial, freshwater or marine systems (Wotton, 1992; Murakami and Nakano, 2000; Strong *et al.*, 2000; Sanz, 2001). Birds play a crucial role for successful pollination as demonstrated by Robertson *et al.* (1999) in New Zealand. Similarly, birds are important seed dispersers and can have profound effects on forest regeneration (Bleher and Böhning-Gaese, 2001). Climate-related changes in the abundance of birds might, therefore, have cascading, far-reaching effects as, vice versa, changes in the ecosystem are expected to cause changes in bird community composition.

Finally, current climate change can, somewhat cynically, be seen as a vast natural experiment. Current ecological theory has advanced to the point where a number of patterns and processes are well understood. With current climate change we are now able to put these theories on test. For example, many studies have shown that the number of migratory bird individuals and species in bird communities is related to the seasonality of their environment. With increasing minimum temperatures and decreasing seasonality it is possible to make clear predictions about the expected changes in bird communities. Within the next years, it will be possible to actually test if our understanding of birds' ranges and communities has predictive power.

REFERENCES

Ainley, D., Wilson, P. and Fraser, W.R. (1994) In: *Impacts of Climate Change on Wildlife* (Ed. by R.E. Green, M. Harley, M. Spalding and C. Zöckler), pp. 26–27. Royal Society for the Protection of Birds, Sandy.

Bale, J.S., Masters, G.J., Hodkinson, I.D., Awmack, C., Bezemer, T.M., Brown, V.K., Butterfield, J., Buse, A., Coulson, J.C., Farrar, J., Good, J.E.G., Harrington, R., Hartley, S., Jones, T.H., Lindroth, R.L., Press, M.C., Symrnioudis, I., Watt, A.D. and Whittaker, J.B. (2002) *Global Change Biol.* **8**, 1–16.

Bauer, H.-G. and Bezzel, E. (2001) Neubürger. In: *Taschenbuch für Vogelschutz* (Ed. by K. Richarz, E. Bezzel and M. Hormann), pp. 100–115. Aula-Verlag, Wiebelsheim.

Begon, M., Harper, J.L. and Townsend, C.R. (1996) *Ecology*. Blackwell Science Ltd, Oxford.

Benning, T.L., LaPointe, D., Atkinson, C.T. and Vitousek, P.M. (2002) *Proc. Natl Acad. Sci. USA* **99**, 14247–14249.

Berry, P.M., Dawson, T.P., Harrison, P.A. and Pearson, R.G. (2002) *Global Ecol. Biogeogr.* **11**, 453–462.

Berthold, P. (2000) *Vogelzug. 4. Auflage*. Wissenschaftliche Buchgesellschaft, Darmstadt.

Bleher, B. and Böhning-Gaese, K. (2001) *Oecologia* **129**, 385–394.

Bogaert, J., Zhou, L., Tucker, C.J., Myneni, R.B. and Ceulemans, R. (2002) *J. Geophys. Res.-Atmos.* **107**(art. no. 4119).

Böhning-Gaese, K., Taper, M.L. and Brown, J.H. (1993) *Conserv. Biol.* **7**, 76–86.

Brown, J.H. and Lomolino, M.V. (1998) *Biogeography*, 2nd Ed. Sinauer Associates, Inc., Sunderland.

Brown, J.H., Mehlman, D.W. and Stevens, G.C. (1995) *Ecology* **76**, 2028–2043.

Burton, J.F. (1995) *Birds and Climate Change*. Christopher Helm, London.

Cameron, G.N. and Scheel, D. (2001) *J. Mammal.* **82**, 652–680.

Crick, H.Q.P., Dudley, C., Glue, D.E. and Thomson, D.L. (1997) *Nature* **388**, 526.

Currie, D.J. (1991) *Am. Nat.* **137**, 27–49.

Currie, D.J. (2001) *Ecosystems* **4**, 216–225.

Davis, A.J., Jenkinson, L.S., Lawton, J.H., Shorrocks, B. and Wood, S. (1998) *Nature* **391**, 783–786.

Dobzhansky, T. (1950) *Am. Sci.* **38**, 209–221.

Dunlop, N. (2001) *West. Fisheries Mag. Spring*, 11–14.

Easterling, D.R., Horton, B., Jones, P.D., Peterson, T.C., Karl, T.R., Parker, D.E., Salinger, M.J., Razuvayev, V., Plummer, N., Jamason, P. and Folland, C.K. (1997) *Science* **277**, 364–367.

Easterling, D.R., Meehl, G.A., Parmesan, C., Changnon, S.A., Karl, T.R. and Mearns, L.O. (2000) *Science* **289**, 2068–2074.

Erasmus, B.F.N., van Jaarsveld, A.S., Chown, S.L., Kshatriya, M. and Wessels, K.J. (2002) *Global Change Biol.* **8**, 689–693.

Galbraith, H., Jones, R., Park, R., Clough, J., Herrod-Julius, S., Harrington, B. and Page, G. (2002) *Waterbirds* **25**, 173–183.

Gaston, K.J. (Ed.) (1996) *Biodiversity*. Blackwell Science Ltd, Oxford.

Gaston, K.J. (2003) *The Structure and Dynamics of Geographic Ranges*. Oxford University Press, Oxford.

Grabherr, G., Gottfried, M. and Pauli, H. (1994) *Nature* **369**, 448.

Gross, S.J. and Price, T.D. (2000) *J. Biogeogr.* **27**, 869–878.

Hannah, L., Midgley, G.F. and Millar, D. (2002) *Global Ecol. Biogeogr.* **11**, 485–495.

Hawkins, B.A., Porter, E.E. and Diniz, J.A.F. (2003) *Ecology* **84**, 1608–1623.

Hayworth, A.M. and Weathers, W. (1984) *Condor* **86**, 19–26.

Herrera, C.M. (1978) *Auk* **95**, 469–509.

Hill, J.K., Thomas, C.D. and Huntley, B. (1999) *Proc. R Soc. Lond. Series B-Biol. Sci.* **266**, 1197–1206.

Hill, J.K., Thomas, C.D., Fox, R., Telfer, M.G., Willis, S.G., Asher, J. and Huntley, B. (2002) *Proc. R. Soc. Lond. Series B-Biol. Sci.* **269**, 2163–2171.

Hofer, U., Bersier, L.F. and Borcard, D. (1999) *Ecology* **80**, 976–988.

Hulme, P.E. (2003) *Oryx* **37**, 178–193.

Hurlbert, A.H. and Haskell, J.P. (2003) *Am. Nat.* **161**, 83–97.

Husak, M.S. and Maxwell, T.C. (2000) *Texas J. Sci.* **52**, 275–284.

Intergovernmental Panel on Climate Change (IPCC). (2001) *Third Assessment Report of the Intergovernmental Panel on Climate Change IPCC (WG I and II)*. Cambridge University Press, Cambridge.

Johnson, N.K. (1994) *Stud. Avian Biol.* **15**, 27–44.

Jones, P.D., New, M., Parker, D.E., Martin, S. and Gregor, I.G. (1999) *Rev. Geophys.* **37**, 173–199.

Kalela, O. (1949) *Bird-Banding* **20**, 77–103.

Kalela, O. (1952) *Fennia* **75**, 38–57.

Kirschbaum, M.U.F. (2000) *Tree Physiol.* **20**, 309–322.

Kont, A., Jaagus, J. and Aunap, R. (2003) *Global Planet. Change* **36**, 1–15.

Konvicka, M., Maradova, M., Benes, J., Fric, Z. and Kepka, P. (2003) *Global Ecol. Biogeogr.* **12**, 403–410.

Lemoine, N. and Böhning-Gaese, K. (2003) *Conserv. Biol.* **17**, 577–586.

Lennon, J.J., Greenwood, J.J.D. and Turner, J.R.G. (2000) *J. Anim. Ecol.* **69**, 581–598.

MacArthur, R.H. (1972) *Geographical Ecology. Patterns in the Distribution of Species.* Harper and Row, New York.

McCarty, J.P. (2001) *Conserv. Biol.* **15**, 320–331.

McDonald, K.A. and Brown, J.H. (1992) *Conserv. Biol.* **6**, 409–415.

Mehlmann, D.W. (1997) *Ecol. Appl.* **7**, 614–624.

Midgley, G.F., Hannah, L., Millar, D., Rutherford, M.C. and Powrie, L.W. (2002) *Global Ecol. Biogeogr.* **11**, 445–451.

Moreira, F. (1997) *Estuar. Coast. Shelf Sci.* **44**, 67–78.

Moss, R. (1986) *Ibis* **128**, 65–72.

Murakami, M. and Nakano, S. (2000) *Proc. R. Soc. Lond. Series B-Biol. Sci.* **267**, 1597–1601.

Newton, I. (2003) *The Speciation and Biogeography of Birds.* Academic Press, London.

Newton, I. and Dale, L. (1996a) *J. Anim. Ecol.* **65**, 137–146.

Newton, I. and Dale, L. (1996b) *Auk* **113**, 626–635.

Parmesan, C. (1996) *Nature* **382**, 765–766.

Parmesan, C. and Yohe, G. (2003) *Nature* **421**, 37–42.

Parmesan, C., Ryrholm, N., Stefanescu, C., Hill, J.K., Thomas, C.D., Descimon, H., Huntley, B., Kaila, L., Kullberg, J., Tammaru, T., Tennent, W.J., Thomas, J.A. and Warren, M. (1999) *Nature* **399**, 579–583.

Parmesan, C., Root, T.L. and Willig, M.R. (2000) *Bull. Am. Meteorol. Soc.* **81**, 443–450.

Pearson, R.G. and Dawson, T.P. (2003) *Global Ecol. Biogeogr.* **12**, 361–371.

Peterson, A.T. (2003a) *Global Change Biol.* **9**, 647–655.

Peterson, A.T. (2003b) *Southwest. Nat.* **48**, 289–292.

Peterson, A.T., Sánchez-Cordero, V., Soberón, J., Bartley, J., Buddemeier, R.W. and Navarro-Sigüenza, A.G. (2001) *Ecol. Model.* **144**, 21–30.

Peterson, A.T., Ortega-Huerta, M.A., Bartley, J., Sánchez-Cordero, V., Soberón, J., Buddemeier, R.H. and Stockwell, D.R.B. (2002) *Nature* **416**, 626–629.

Pounds, J.A., Fogden, M.P.L. and Campbell, J.H. (1999) *Nature* **398**, 611–615.

Rabenold, K.N. (1979) *Am. Nat.* **114**, 275–286.

Rabenold, K.N. (1993) In: *Latitudinal Gradients in Avian Species Diversity and the Role of Long-Distance Migration* (Ed. by D.M. Power), Curr. Ornithol., Vol. 10. pp. 247–274. Plenum Press, New York.

Repasky, R.R. (1991) *Ecology* **72**, 2274–2285.

Robertson, A.W., Kelly, D., Ladley, J.J. and Sparrow, A.D. (1999) *Conserv. Biol.* **13**, 499–508.

Rodriguez, J.P. (2002) *Ecol. Appl.* **12**, 238–248.

Root, T.L. (1988) *Ecology* **69**, 330–339.

Root, T.L. and Schneider, S.H. (1993) *Conserv. Biol.* **7**, 256–278.

Root, T.L., Price, J.T., Hall, K.R., Schneider, S.H., Rosenzweig, C. and Pounds, J.A. (2003) *Nature* **421**, 57–60.

Sanz, J.J. (2001) *Ecol. Res.* **16**, 387–394.

Saunders, D.A. and Curry, P.J. (1990) *Proc. Ecol. Soc. Aust.* **16**, 303–321.

Schwartz, M.W. (1992) *For. Chron.* **68**, 462–471.

Shine, R., Barrott, E.G. and Elphick, M.J. (2002) *Ecology* **83**, 2808–2815.

Stapp, P., Polis, G.A. and Piñero, F.S. (1999) *Nature* **401**, 467–469.

Stockwell, D.R.B. and Noble, I.R. (1992) *Math. Comp. Simulation* **33**, 385–390.

Stockwell, D.R.B. and Peters, D. (1999) *Int. J. Geogr. Inform. Sci.* **13**, 143–158.

Strong, A.M., Sherry, T.W. and Holmes, R.T. (2000) *Oecologia* **125**, 370–379.

Thomas, C.D. and Lennon, J.J. (1999) *Nature* **399**, 213.

Turner, J.R.G., Lennon, J.J. and Lawrenson, J.A. (1988) *Nature* **335**, 539–541.

Van Rensburg, B.J., Chown, S.L. and Gaston, K.J. (2002) *Am. Nat.* **159**, 566–577.

Villard, M.A. and Maurer, B.A. (1996) *Ecology* **77**, 59–68.

Visser, M.E., van Noordwijk, A.J., Tinbergen, J.M. and Lessells, C.M. (1998) *Proc. R. Soc. Lond. Series B-Biol. Sci.* **265**, 1867–1870.

Walther, G.-R., Burga, C.A., and Edwards, P.J. (Eds.) (2001) *"Fingerprints" of Climate Change*. Kluwer Academic/Plenum Publishers, New York.

Warren, M.S., Hill, J.K., Thomas, J.A., Asher, J., Fox, R., Huntley, B., Roy, D.B., Telfer, M.G., Jeffcoate, S., Harding, P., Jeffcoate, G., Willis, S.G., Greatorex-Davies, J.N., Moss, D. and Thomas, C.D. (2001) *Nature* **414**, 65–69.

Wotton, J.T. (1992) *Ecology* **73**, 981–991.

Zöckler, C. and Lysenko, I. (1994) In: *Impacts of Climate Change on Wildlife* (Ed. by R.E. Green, M. Harley, M. Spalding and C. Zöckler), pp. 20–25. Royal Society for the Protection of Birds, Sandy.

The Challenge of Future Research on Climate Change and Avian Biology

ANDERS P. MØLLER,* PETER BERTHOLD
AND WOLFGANG FIEDLER

I. SUMMARY

Current knowledge of the effects of climate change on avian biology is mainly based on studies of passerines from the temperate zone of the northern hemisphere. We need more studies from other climate zones and of species belonging to orders other than the passerines to redress this imbalance. The frequency of extreme environmental conditions is predicted to increase, but the impact of such extreme conditions is relatively little studied, because such events by definition are rare. Intraspecific and interspecific interactions may change in frequency and intensity because the abundance of conspecifics may change, but also because the abundance of parasites and predators may change. The relative role of phenotypic plasticity and micro-evolutionary change as mechanisms allowing phenotypic change in response to environmental conditions needs to be assessed, and the ecological factors predicting interspecific differences in responses to climate change need to be identified. Latitudinal gradients in phenotype provide unexploited natural model systems that can be used to gain insights into the effects of climate change on birds. We suggest that large-scale

---

E-mail address: amoller@snv.jussieu.fr (A.P. Møller)

ADVANCES IN ECOLOGICAL RESEARCH, VOL. 35
0065-2504/04 $35.00 DOI 10.1016/S0065-2504(04)35011-7

studies of a number of model species along such latitudinal gradients will provide a better understanding of the processes that have resulted in adaptation of birds to local environmental conditions.

II. INTRODUCTION

This book provides the most extensive review of the effects of climate and climate change on the biology of a single class of organisms; the birds. This group is particularly well suited for such a treatment because an extraordinary amount is known about all aspects of their biology. More is known about this class of animals than any other class.

Despite these achievements we must still admit that we know very little about the details in which climate affects organisms. A summary of studies that are particularly in need of being conducted to allow us to gain better knowledge of the effects of climate change on free-living birds is given in Table 1.

III. THE EFFECTS OF CLIMATE ON AVIAN BIOLOGY

Our current knowledge of the effects of climate change on avian biology is almost entirely restricted to studies in the temperate zone of the northern hemisphere, in particular the European and North American parts of this zone. We clearly need to know more about the effects of climate change in other climate zones. Since the effects of climate change are predicted to be particularly severe at high latitudes (Houghton et al., 2001), we need more studies from Arctic regions. In addition, we need more information about the impact of climate change in subtropical and tropical regions. Very few studies have addressed the impact of climate change in the southern hemisphere and this lack of knowledge must clearly be redressed.

We know very little about the importance of local versus global environmental conditions on free-living organisms. Local weather conditions are by necessity related to continental or global weather phenomena. Some studies have shown clear correlations between indicators of continental weather conditions such as the North Atlantic Oscillation or the El Niño-Southern Oscillation and biological phenomena (e.g., Marra et al., 1998; Møller, 2002; Przybylo et al., 2000; Sæther et al., 2000; Sillett et al., 2000; Hüppop and Hüppop, 2003; Jenni and Kéry, 2003; Møller and Erritzøe, 2003). Other studies have found strong relationships between local weather conditions and biological phenomena such as population dynamics and intensity of selection (e.g., Sæther et al., 2000; Both and Visser, 2001). However, we lack proper knowledge of the relative efficiency of these two different measures of weather phenomena as predictors of biological phenomena, using the same data sets. Is effect size greater for general weather indicators or for local weather indicators?

Table 1 List of areas of research where further investigation of the effects of climate change on birds may be particularly rewarding

Problem	Studies required
Geographical distribution of studies	Studies from other regions than northern temperate zones
Taxonomic distribution of studies	Studies of other orders than passerines
Spatial scale of weather conditions	The relative role of local and global weather systems
Specific weather conditions	Detailed studies of responses of individual birds to specific weather variables
Extreme weather conditions	The effects of extreme weather conditions
Scientific approach	More experiments are needed
Interspecific interactions	Changing impact of predators and parasites
Intraspecific interactions	Changing importance of density-dependence
Effects of climate change on phenotypic plasticity	Degree of phenotypic plasticity under different environmental conditions
Phenotypic and evolutionary responses to climate change	Relative importance of phenotypic plasticity and micro-evolutionary adaptation
Trait-specific responses to climate change	Which traits respond to climate change and why?
Interspecific differences in response to climate change	Which species respond to climate change and why?
Complex annual cycles	The relative role of environmental conditions during breeding, migration and wintering for adaptation
Latitudinal clines	Latitudinal gradients in phenotypic characters and their maintenance: Collaboration of amateurs and professionals
Historical changes in phenotype	Use of museum collections to investigate temporal changes in phenotype in response to climate change
Heterogeneity in responses to climate change	Age and sex differences in response to climate change

A clear prediction arising from current studies of climate change is that the variance in weather will increase as the climate becomes warmer (Houghton *et al.*, 2001). Effects of extreme weather phenomena may be much more severe than effects of small changes in average conditions. For example, one of us (PB) has noticed that blackcaps *Sylvia atricapilla* did not breed at all during a period of one and a half weeks in the summer 1995 when extremely heavy rain prevailed in Southern Germany. Likewise, in the extremely warm late summer of 2003 in Europe barn swallows *Hirundo rustica* had severely reduced reproductive success in their second broods and their offspring were emaciated and had extremely weak T-cell mediated immune responses compared to first broods, or to second broods during the previous years (A. P. Møller *et al.*, unpublished data). Extreme weather phenomena are by definition rare, and therefore difficult to study, but they may have particularly severe impact on populations of free-living organisms.

Most research on climate change is entirely descriptive in nature, making it impossible to assign cause and effect. Experimentation is indeed possible as shown by some classical studies. For example, Nager and van Noordwijk (1992, 1996) experimentally heated nest boxes of the great tit *Parus major* during the laying phase, thereby mimicking the situation when early spring temperatures are increasing. They found that such an increase in temperatures caused an increase in reproductive success because laying dates were affected by changing environmental conditions early during the breeding season. We need much more experimental approaches to this subject in order to be able to draw firm conclusions.

Most research on the effects of climate change is based on mere descriptions and post hoc interpretations. This situation is only natural given that the phenomenon has only been the focus of study for less than two decades. However, we do need to move beyond mere description by adopting a modelling approach and testing quantitative predictions.

Intraspecific and interspecific interactions are likely to be affected by climate change although our ignorance of these aspects of the population biology of birds is almost complete. Parasites and predators are important causes of mortality and reduced reproductive potential in birds. As environmental conditions at high latitudes ameliorate due to climate change, we can expect that population density of many species of birds increases. We can also expect that the part of the annual cycle where parasites are active in exploiting their hosts will increase, thereby causing an increase in parasite-induced natural selection (Mouritsen and Poulin, 2002). However, the impact of parasites on host populations is difficult to assess because the ability of hosts to defend themselves is also likely to change in response to climate change (Møller, 2002; Møller and Erritzøe, 2003). Predation is often density-dependent, and higher population density of prey may feed back on the density of predator populations. High population densities are also likely to increase the intensity of intraspecific competition through effects of density-dependence on demographic variables (Sæther *et al.*, 2000, Sæther *et al.*, 2004, this volume). Whether such effects of density-dependence will have a net effect on population sizes remains to be determined.

IV. PHENOTYPIC PLASTICITY OR MICRO-EVOLUTIONARY CHANGE

The relative roles of phenotypic plasticity and micro-evolution in allowing birds to adjust to changing environmental conditions are still unknown. We know from artificial selection experiments that birds can respond very rapidly to changes in environmental conditions by adjusting their migration to environmental conditions (Berthold et al., 1990; Pulido et al., 1996). Field studies have also shown that birds have adapted to changes in environmental conditions (e.g., Berthold et al., 1992). However, we know very little about the extent of phenotypic plasticity and how the degree of phenotypic plasticity depends on environmental conditions (Møller and Merilä, 2004, this volume; Coppack and Pulido, 2004, this volume; Pulido and Berthold, 2004, this volume). Studies of birds could readily investigate these phenomena.

Lack of response to selection due to changes in environmental conditions may have multiple causes. These include negative genetic correlations, lack of potential cues that might allow a response to climate change, and long distance migration constraining response to climate change. For example, studies of migration and arrival dates have shown clear evidence of an overall response to climate change (Lehikoinen et al., 2004, this volume), although not all populations or species have shown such effects. Which are the phenotypic traits that respond to climate change, and which do not? Why are there such differences among traits, and why do some species respond while others do not? Comparative analyses can contribute significantly to an improvement in our understanding of the features that prevent or limit adaptation to changes in environmental conditions.

Responses to climate change have generally been assessed as the mean response at the population level. Such an approach invariable disregards heterogeneity in response due to age- and sex-differences in response to climate change. A prime example of this problem is sex differences in arrival date in migratory birds. Males generally arrive before females and such protandry is thought to arise as a consequence of sex differences in the costs and benefits of early arrival (Morbey and Ydenberg, 2001). While competitively superior males benefit from early arrival through increased gains in terms of acquisition of mates and/or territories, females gain less from early arrival. Previous studies of the effects of climate change on bird migration have consistently investigated first arrival dates of birds, which invariably concerns only males (Lehikoinen et al., 2004, this volume). There are good theoretical reasons to expect that individuals of the two sexes should respond differently to climate change because of sex differences in costs and benefits of early arrival. A long-term study of arrival dates of male and female barn swallows revealed that only males responded to an amelioration of climatic conditions during spring migration, while females did not (Møller, 2004a). Therefore, the sex difference in arrival date increased as a consequence of climate change, thereby effectively preventing a change in breeding date because females do not arrive earlier now than they did more than 30 years ago. This study clearly shows that it is important to consider different sex and age classes when investigating the effects of and response to climate change.

V. COMPLEX ANNUAL CYCLES

Migratory birds are special by having complex annual cycles because they spend part of the year in the breeding areas, another part during migration and yet another part in the winter quarters. These different stages of the annual cycle may be physically located many 1000 km apart. For example, the Arctic tern *Sterna paradisaea* breeds in the temperate and Arctic regions of the northern hemisphere, migrates across the equator in the Atlantic and then spends the "winter" in Antarctic waters. Numerous examples of less extreme migrations are known (Berthold, 2001). Such species must adapt to the environmental conditions in all the different parts of the annual range, making the constraints to adaptation discussed above even more challenging than those for resident species that spend their entire annual cycle within a specific area.

Avian biologists have started to investigate how effects of environmental conditions in disparate parts of the annual range have carry-over consequences from one part of the range to another. For example, studies have now shown clear evidence of carry-over effects from the winter quarters to the breeding grounds (Marra et al., 1998; Møller and Hobson, 2004; Norris et al., 2004; Saino et al., 2004a,b), and from stop-over sites during migration to the breeding grounds (Møller, 2004b). Climate change also has important implications for population dynamics by carry-over effects from the wintering period to the breeding grounds (Sillett et al., 2000; Saino et al., 2004a). The importance of selection and adaptation in each of these disparate parts of the annual range still needs to be assessed.

VI. LATITUDINAL GRADIENTS AS MODEL SYSTEMS

Many species of birds have enormous ranges of distribution from the tropics over the subtropics to the temperate and the Arctic climate zones. Adaptations to these disparate climatic conditions have only been studied to a limited extent. Latitudinal gradients in phenotypes may provide us with model systems that can be used to achieve a better understanding of the effects of climate change.

Organisms with a large latitudinal range of distribution are particularly well suited for studies of the effects of climate change on ecological and evolutionary processes. For example, extensive studies of common hole nesting passerines such as the great tit *Parus major* and the pied flycatcher *Ficedula hypoleuca* have revealed that population size is more strongly influenced by the effects of the North Atlantic Oscillation (a large-scale weather indicator based on a measure of pressure difference between Iceland and the Azores) at high latitudes (Sæther *et al.*, 2003). Likewise, studies of demographic variables also show strong relationships with latitude (Sanz, 2002, 2003; Sanz *et al.*, 2003; Visser *et al.*, 2003). However, there are overall relatively few studies of the effects of climate change based on investigations of latitudinal gradients. This is all the more surprising given that such gradients are likely to reflect the likely changes in phenotypic variables as the climate changes.

Such studies of latitudinal gradients may be particularly rewarding for species that are currently investigated by a large number of research groups: barn swallow, tree swallow *Tachycineta bicolor*, great tit, blue tit *Parus caeruleus*, pied flycatcher and several others. Such studies would benefit from collaboration between amateurs and professionals. Neither can successfully achieve the goal of understanding so complex phenomena on their own.

Museum collections contain vast numbers of specimens of birds, but also of clutches of birds. Most of these were collected in the 19th and the early 20th centuries. Surprisingly, there have been very few attempts to use these collections to investigate effects of climate change on avian biology. Collections can be used to investigate changes in the distribution of genotypes across the distributional range in an attempt to test whether changes in climate are associated with northward changes in the distribution of specific genotypes. These collections can also be used to investigate changes in phenotype of juveniles and adults and phenotype of eggs over time. In particular, more than 30 species of birds have changed status from migrants to residents in Northern Germany and Southern Scandinavia since the mid 1800s. This provides a unique opportunity to investigate how changes in migratory habits are associated with changes in morphology and life history traits such as clutch size and egg size during a period of climate change. Many species from the same area that have not changed migratory status during the same period could serve as "controls" for general temporal changes in phenotype unrelated to changes in migratory status.

VII. CONCLUSIONS

The first study addressing the potential consequences of climate change on the biology of birds only dates back less than 15 years (Berthold, 1991). Since then we have gained tremendous knowledge in this field, and every week provides large amounts of information through a steady stream of new publications.

We consider that studies of climate change comprise a unique opportunity to study the adaptation of organisms to their changing environments. Birds are particularly suitable for this study because individuals can readily be followed throughout their entire lives and across generations. This allows for individual-based studies of physiology, ecology and evolution. Numerous aspects of how birds adapt to their environment are still in need of study. We have listed a number of such pressing questions in Table 1. Hopefully, this book will contribute to this endeavour by clearly and critically summarising the field while carefully assessing the limits to our current knowledge. We gratefully acknowledge the efforts of our contributors, and we feel sure that their reviews will remain as summaries of the state of the art for the coming decade, when climate is predicted to change at an even faster rate than it has during the last century.

REFERENCES

Berthold, P. (1991) *Acta XX Congr. Int. Orn.*, 780–786.
Berthold, P. (2001) *Bird Migration*, 2nd Ed. Oxford University Press, Oxford.
Berthold, P., Mohr, G. and Querner, U. (1990) *J. Orn.* **131**, 33–45.
Berthold, P., Helbig, A.J., Mohr, G. and Querner, U. (1992) *Nature* **360**, 668–670.
Both, C. and Visser, M.E. (2001) *Nature* **411**, 296–298.
Houghton, J.T., Ding, Y., Griggs, D.J., Noguer, M., Van der Linden, P.J. and Xiasou, D. (2001) *Climate Change 2001*. Cambridge University Press, Cambridge.
Hüppop, O. and Hüppop, K. (2003) *Proc. R. Soc. Lond. B* **270**, 233–240.
Jenni, L. and Kéry, M. (2003) *Proc. R. Soc. Lond. B* **270**, 1467–1471.
Marra, P.P., Hobson, K.A. and Holmes, R.T. (1998) *Science* **282**, 1884–1886.
Møller, A.P. (2002) *J. Anim. Ecol.* **71**, 201–210.
Møller, A.P. (2004a) *Global Change Biol.* (in press).
Møller, A.P. (2004b) *J. Evol. Biol.* (in press).
Møller, A.P. and Erritzøe, J. (2003) *Oecologia* **137**, 621–626.
Møller, A.P. and Hobson, K.A (2004) *Proc. R. Soc. Lond. B* **271**, 1355–1362.
Morbey, Y.E. and Ydenberg, R.C. (2001) *Ecol. Lett.* **4**, 663–673.
Mouritsen, K.N. and Poulin, R. (2002) *Oikos* **97**, 462–468.
Nager, R.G. and van Noordwijk, A.J. (1992) *Proc. R. Soc. Lond. B* **249**, 259–263.
Nager, R.G. and van Noordwijk, A.J. (1996) *Am. Nat.* **146**, 454–474.
Norris, D.R., Marra, P.P., Kyser, T.K., Sherry, T.W. and Ratcliffe, L.M. (2004) *Proc. R. Soc. Lond. B* **271**, 59–64.
Przybylo, R., Sheldon, B.C. and Merilä, J. (2000) *J. Anim. Ecol.* **69**, 395–403.
Pulido, F., Berthold, P. and van Noordwijk, A.J. (1996) *Proc. Natl. Acad. Sci. USA* **93**, 14642–14647.
Sæther, B.E., Tufto, J., Engen, S., Jerstad, K., Røstad, O.W. and Skåtan, J.E. (2000) *Science* **287**, 854–856.

Sæther, B.-E., Engen, S., Møller, A.P., Matthysen, E., Adriaensen, F., Fiedler, W., Leivits, A., Lambrechts, M.M., Visser, M., Anker-Nilssen, T., Both, C., Dhondt, A., McCleery, R.H., McMeeking, J., Potti, J., Røstad, O.W. and Thomson, D. (2003) *Proc. R. Soc. Lond. B* **270**, 2397–2404.

Saino, N., Szép, T., Romano, M., Rubolini, D. and Møller, A.P. (2004a) *Ecol. Lett.* **7**, 21–25.

Saino, N., Szép, T., Ambrosini, R., Romano, M. and Møller, A.P. (2004b) *Proc. R. Soc. Lond. B* **271**, 681–686.

Sanz, J.J. (2002) *Global Change Biol.* **8**, 1–14.

Sanz, J.J. (2003) *Ecography* **26**, 45–50.

Sanz, J.J., Potti, J., Moreno, J., Merino, S. and Frías, O. (2003) *Global Change Biol.* **9**, 1–12.

Sillett, T.S., Holmes, R.T. and Sherry, T.W. (2000) *Science* **288**, 2040–2042.

Visser, M.E., Adriaensen, F., van Balen, J., Blondel, J., Dhondt, A.A., van Dongen, S., du Feu, C., Ivankina, E.V., Kerimov, A.B. and de Laet, J. (2003) *Proc. R. Soc. Lond. B* **270**, 367–372.

Subject Index

Advances in Ecological Research
Volume 1–35

Cumulative List of Titles

Aerial heavy metal pollution and terrestrial ecosystems, **11**, 218

Age determination and growth of Baikal seals (*Phoca sibirica*), **31**, 449

Age-related decline in forest productivity: pattern and process, **27**, 213

Analysis and interpretation of long-term studies investigating responses to climate change, **35**, 111

Analysis of processes involved in the natural control of insects, **2**, 1

Ancient Lake Pennon and its endemic molluscan faun (Central Europe; Mio-Pliocene), **31**, 463

Ant-plant-homopteran interactions, **16**, 53

Arrival and departure dates, **35**, 1

The benthic invertebrates of Lake Khubsugul, Mongolia, **31**, 97

Biogeography and species diversity of diatoms in the northern basin of Lake Tanganyika, **31**, 115

Biological strategies of nutrient cycling in soil systems, **13**, 1

Bray-Curtis ordination: an effective strategy for analysis of multivariate ecological data, **14**, 1

Breeding dates and reproductive performance, **35**, 69

Can a general hypothesis explain population cycles of forest lepidoptera?, **18**, 179

Carbon allocation in trees; a review of concepts for modeling, **25**, 60

Catchment properties and the transport of major elements to estuaries, **29**, 1

Coevolution of mycorrhizal symbionts and their hosts to metal-contaminated environment, **30**, 69

Conservation of the endemic cichlid fishes of Lake Tanganyika; implications from population-level studies based on mitochondrial DNA, **31**, 539

The cost of living: field metabolic rates of small mammals, **30**, 177

A century of evolution in *Spartina anglica*, **21**, 1

The challenge of future research on climate change and avian biology, **35**, 237

Climate influences on avian population dynamics, **35**, 185

The climatic response to greenhouse gases, **22**, 1

Communities of parasitoids associated with leafhoppers and planthoppers in Europe, **17**, 282